THE TRUE TRANSATLANTIC
SUPER-LINERS

IMPERATOR

THE TRUE TRANSATLANTIC SUPER-LINERS

AN EXCLUSIVE CLASS FROM IMPERATOR TO QM2

DAVID L. WILLIAMS AND RICHARD P. DE KERBRECH

The History Press

Cover illustrations:

Front: '168,800 gross tons of Super-Liners'. A painting depicting Albert Ballin's big three in Southampton's Ocean Dock on 23 July 1930. *Leviathan* (left) arrived first and bagged the preferred inner berth, followed by *Majestic* (centre). *Berengaria*, on the opposite side of the dock, is being guided into her berth. (© Stephen J. Card AFNI)

Back, top: Queen Mary 2 seen in mid-Atlantic from *Queen Elizabeth 2* when they crossed in tandem in October 2008, a unique event in the story of the transatlantic super-liners; *bottom*: Just days later, on 16 October 2008, *Queen Elizabeth 2* is seen from the Brooklyn Cruise Terminal, making her final departure from New York and ending her forty-year service career. (Both Bill Burt)

Frontispiece: The newly commissioned *Imperator* from the port bow. (US Library of Congress)

First published 2024

The History Press
97 St George's Place
Cheltenham
Gloucestershire
GL50 3QB
www.thehistorypress.co.uk

British Library Cataloguing in Publication Data.
A catalogue record for this book is available from the British Library.

ISBN 978 1 80399 551 9

Typesetting and origination by The History Press
Printed and bound in India by Thomson Press India Ltd

CONTENTS

INTRODUCTION

SUPER-LINERS: DEFINITION AND DISTINCTION

In order for the reader to appreciate the precise limitations applied to the book's scope, it is necessary, first, to elucidate the meanings of terminology relevant to the categorisation of 'super', to make certain distinctions, and to dispel some misconceptions.

So, what exactly is a 'super-liner'? Clearly, one would expect them to stand apart from other passenger ships, a distinction that needed to be quantifiable to avoid any ambiguity. The designation was in fact retrospective because for the first twenty-three years of the existence of liners so categorised, the term 'super-liner' had not been coined. But this was addressed in 1938 when George C. Gaede, the Passenger Traffic Manager of American Export Lines, presented a paper entitled 'American Super-Liners – They Will Pay!' to the Society of Naval Architects & Marine Engineers (SNAME) in New York. His exact words were:

> The word 'super-liner' was created to designate the first ship of this character, which was the *Imperator*, later the *Berengaria*, built in 1912, of a gross tonnage of 52,226 and an average speed of 22.49 knots, for rating purposes in the Conference*. The previous largest ship of the Hamburg Amerika Line, the *Kaiserin Auguste Viktoria*, built in 1905, had a tonnage of 25,160 and a speed of 16.97 knots. The leap from 25,160 to 52,226 tons suggested the word 'super-liner'. Later, the term was applied to all [passenger] ships of 50,000 tons or larger. In the trade, therefore, the term has a very definite meaning, although now and then it is carelessly applied to ships of a much smaller tonnage.

* North Atlantic Steamship Lines Association Conference.

Thus, super-liners were, and are, passenger liners having a gross registered tonnage of 50,000 and higher, and invariably of around 900ft and greater in length.

It may be thought that other great Atlantic ocean liners deserve to be included here as worthy of the epithet 'super'. One immediately thinks of *Olympic*, *Titanic* and their intended fleet mate *Britannic*, along with *Aquitania*, *Conte di Savoia* and the later *Michelangelo* and *Raffaello*. All of them were immense and impressive but, as can be seen in the table below, without exception they fell outside the defined parameters.

	Gross Tonnage	Length (ft)
Olympic (White Star)	46,440	882
Titanic (White Star)	46,329	883
Britannic (White Star)	48,160	903
Aquitania (Cunard)	45,645	901
Conte di Savoia (Italia)	48,502	815
Michelangelo (Italia)	45,991	906
Raffaello (Italia)	45,935	905

Of the ships described herein, at the bottom of the super-liner roster is the Italian-flag *Rex*. It can be seen that, at 51,730 gross tons she was only just over 3,000 tons larger than her consort *Conte di Savoia*, while she was shorter at 880ft than all the other ships listed in the table above. But *Rex*, like others of the super-liners, was also a racer, another of the distinctive credentials that applied to certain of them, and she held the accolade of being the fastest ship in the world. It should be stressed that six of the thirteen liners of the 'super' class were at some point holders of the Atlantic Blue Riband in recognition of their speed honours. Indeed, *United States*, as a passenger liner, is rightly regarded as still being the true holder of the transatlantic speed record, a distinction she will now retain in perpetuity. Of the remaining seven super-liners, four of them that did not actually secure record passage times, were nonetheless certainly capable of achieving the speeds that did result in securing the acclamation of being the fastest in the world.

At the opposite end of the super-liners scale is the mighty *Queen Mary 2*, whose presence here might also be regarded as open to question as she now spends much of her time engaged in cruising. Her right to be among the ships of this category is, however, without dispute for she is unquestionably a true liner. She was conceived to be, like her predecessor *Queen Elizabeth 2*, a dual-purpose vessel intended to maintain a regular transatlantic schedule for a significant part of every year. Cunard made this abundantly clear in its press releases prior to and during her construction. It was stated that she was 'built to keep alive the traditions of the great ocean liners' and that she was 'in all respects a luxury, express passenger liner'. Certain concessions were made in her design to support the cruising dimension of her operational blend but she exhibits many of the features that characterise the true ocean liner and still, some twenty years after she first entered service, she continues to make at least one round-trip Atlantic crossing every month between April and November each year.

There is another distinction that needs to be made, not only to explain an absence from these pages of certain ships that some may feel should not be excluded but also to dispel any perception that the super-liners have somehow had their magnitude and superior standing dwarfed or diminished by the immense cruise ships that now dominate the passenger shipping scene.

The fact is that some of these cruise ships do make occasional transatlantic voyages, for positioning purposes, as part of long-distance cruises or to return to shipyards for routine overhauls. Clouding the picture further is the widespread use nowadays of the term 'cruise liner' when referring to modern, dedicated cruise vessels. But this expression is erroneous as it is a contradiction in terms. In reality there is no such thing as a 'cruise liner'. There are ocean passenger liners and there are cruise ships, which are two distinct vessel paradigms. This book is concerned with the greatest and most illustrious of the former category, the magnificent super-liners that have operated the scheduled transatlantic run. The distinction between these two vessel types is important, especially as modern cruise ships are viewed as an evolved form of the passenger liner of the past and because their most enormous examples are regarded as having eclipsed their predecessors. These categories require greater elucidation, therefore, as a prelude to the stories of the Atlantic super-liners.

In broad terms an ocean liner is a passenger ship operating a single regular service on a fixed transoceanic route or 'line'. The routes to and from New York were once referred to as the 'Atlantic Ferry' for this reason. In many cases, speed was important, both to reduce passage times and to maintain sailing schedules. Accommodation was divided into classes, with the most sophisticated and luxurious in the top category. Conversely, passengers in the lowest class, particularly in the days of steerage, were provided with basic, sometimes spartan, accommodation conditions. Ocean liners invariably carried a quantity of cargo besides passengers, and most were distinguished by having long open foredecks equipped with hatch gear and derricks on their masts. They mostly had circuitous promenade decks and open boat decks with the lifeboats stowed high up. For the express ships, hulls were of a finer form with a high beam-to-length ratio.

In contrast, cruise ships are passenger-carrying vessels that follow various roaming itineraries calling at different tourist ports, conveying passengers who are travelling solely for pleasure on vacation. Their itineraries are not fixed and may be adjusted from voyage to voyage or season to season. They never repeatedly follow a point-to-point route. Typically, their accommodation is elegant and generally of a high standard of appointment and décor for all passengers in a broadly single class. Increasingly, cruise ships are distinguished by their lofty superstructure in order to provide a high ratio of outside balcony staterooms. Thus, with the accommodation placed high up, promenade decks have for the most part been eliminated and lifeboat muster stations have been placed low down within the hull side. This latter change is also a consequence of a SOLAS regulation requiring lifeboats to be at a limited height above the water level. Cruise ships tend to have a broader beam relative to their length (low beam to length ratio) partly in order to aid stability as they also have a shallower draught to permit entry into ports with restricted bottom clearance. They do not generally require the capability of high speed.

The largest cruise vessels are now referred to as mega-cruise ships that, in terms of their tonnage capacity, are those that measure above 100,000 gross tons. In much the same way as the commissioning of *Imperator* occasioned the designation 'super-liner', so the emergence in 1996 of the 101,353-gross-ton *Carnival Destiny* gave rise to the expression mega-cruise-ship, which has been applied to all vessels of this size that followed.

On paper, some of these behemoths are twice or more times the size in gross tonnage terms of the greatest passenger liners, but that can be misleading. Gross Registered Tonnage (GRT) and the modern Gross Tonnage (GT) are measures of volumetric capacity, not of weight, which is displacement tonnage, and it does not necessarily have a bearing on vessels' physical linear dimensions. The unitary value of a Gross Registered Ton was (and still is) a volume of enclosed space of 100 cubic feet, relating to all internal spaces of a ship. Those enclosed spaces that are revenue-earning constitute a ship's Net Registered Tonnage or Net Tonnage upon which its registration costs and port dues are based.

From this understanding of gross tonnage it is possible to determine more distinctions between ocean liners and cruise ships and evaluate how gross tonnage can give a misleading impression of comparative size. Unlike linear dimensions, because gross tonnage is a measure of capacity rather than weight, it can give an impression of size that is not necessarily consistent with physical appearance. As stated, cruise ships often have loftier structures and may be broader across the beam but, in terms of overall length, relatively few of them have longer hulls than those of the super-liners. As an example, *Carnival Destiny*, at just over 100,000 gross tons, was 892ft in length, shorter than all the super-liners with the exception of *Rex*. In the same way, because the super-liners were deeper draught with more of the hull below the waterline, their depth measurements are less different from those of mega-cruise-ships than might be thought.

A point to bear in mind is that gross tonnage can be increased without altering a ship's physical dimensions by the simple expedient of enclosing more space. It could also be achieved by the application of different rules or by adopting contrived methods of calculation. This could have both benefits and disadvantages. In the days of the super-liners there was a certain marketable kudos to not only being the fastest ship but also the largest, and this was reflected in rather arbitrary tonnage changes. For example, *Leviathan*, which measured 54,280 gross tons as *Vaterland*, was re-rated as 59,955 gross tons under the US flag in order for it to be claimed that she was the world's largest ship. The downside of this 'tinkering' with tonnage was that operational costs increase in parallel with the growth of gross tonnage and, in the case of *Leviathan*, her tonnage was subsequently reduced in 1931 to 48,590 gross to trim operating costs. In both instances, no material alteration was made to account for these differences.

A recent case of gross tonnage change without any influence on length or beam dimensions concerns *Queen Mary 2*, which measured 148,528 gross tons when she first entered service in 2004. Modifications carried out in October 2016 that involved the creation of additional stateroom suites, including on her upper deck, resulted in a new tonnage figure of 151,215 gross.

A further contrast between ocean liners and cruise ships that could be made concerns the quantity of their passenger berths, a higher number suggesting a larger ship. Typically, the largest cruise ships now carry 5,000 and more passengers, more or less double that of the great super-liners and, of course, greater tonnage would correlate with having more accommodation spaces. But, as with gross tonnage, there are factors that account for this variance. Already highlighted is the fact that the super-liners carried, beside passengers and mail, a significant quantity of cargo. There is also the matter of the immense physical area allocated to the luxury suites in the deluxe category on certain of the super-liners.

Of greater relevance though, in making comparisons as to the deployment of the interior spaces of these two vessel types, is the expanse occupied by their main engines. The type of motive power is one component – steam turbines compared with gas turbine and electric power – but there is also the size of these respective power units. Achieving the speeds sought in the operation of the super-liners, even allowing for huge improvements in such things as boiler performance, necessitated the installation of immense machinery, trading off one source of revenue (complement size) with another (higher fares for faster crossing times).

In conclusion, something should be said about what brought about the super-liners. In a nutshell, as it has been for the mega-cruise-ships, it was market forces, the demand for greater speed and luxury on the one hand and, on the other, a greater choice of entertainment, recreational and gastronomic options as well as more passenger capacity. Perhaps, too, economy of scale has had a bearing, with the operational costs of one large vessel being less than that of two smaller vessels of half the size.

Having set the scene by demonstrating the exceptional and unique character and size of the true transatlantic super-liners, segregating them from other large liners and distinguishing them from cruise ships, irrespective of their size, it is time to turn to the individual stories of each of these great passenger ships.

ONE

THE *IMPERATOR* CLASS: ALBERT BALLIN'S CRUSHED TEUTONIC DREAM

Towards the end of the Edwardian era, the big players that dominated the transatlantic route planned a three-ship service that would give a regular and reliable timetable. Cunard had both *Lusitania* and *Mauretania* in regular service and were planning the later *Aquitania*, and White Star were working on the *Olympic*-class trio, *Olympic*, *Titanic* and *Britannic*.

Meanwhile, across the North Sea in Germany, the competition was hotting up with Albert Ballin, President of Hamburg Amerika (also known as HAPAG), also planning a three-ship operation. Ballin had joined Hamburg Amerika in 1886, having gained experience running his father's emigration agency, and he was well versed in passengers' requirements and tastes. By 1899 he had risen through the company to become the line's General Manager and Director. Having had past passenger liners named after German royalty, he was also well known to the Kaiser. As a matter of interest, Ballin was also the driving force behind the North Atlantic Steamship Lines Association (Nord-Atlantischer Dampfer Linien Verband). This body sought to represent the interests of all the transatlantic players through its conference to set quotas, rate levels and pool contracts. The intention was to prevent cut-throat competition, to stabilise international passenger traffic and ensure viable and profitable business for all concerned.

Based on previous ships' interior design layout by Charles Mewes in conjunction with César Ritz, Ballin was happy to continue with the new builds and irrevocably committed the Hamburg Amerika Line to the reality that his two British rivals had long since faced, the three-ship service. He had little trouble enlisting the Kaiser's support for such an ambitious scheme. Plans to award the contract to Harland & Wolff of Belfast were cancelled when a new government mail contract stipulated that the ships be constructed in German yards. That they were to be built and launched with clockwork regularity, one a year for three years, is a tribute to the vigour and technical diligence of the Hamburg shipyards. Back in the day, it should be remembered that it was custom and practice for Germany to purchase its passenger ships from British shipyards. So, for an infant industry to embark on such a colossal programme of national importance, let alone complete it, was an extraordinary feat.

IMPERATOR: THE PIONEERING SUPER LINER

The first of HAPAG's trio was laid down at the Vulkan Werke yard, Hamburg, in June 1910 as Yard No. 314.

There were benefits to be derived by HAPAG from starting third after the two British competitors. The ship's design and specifications could incorporate all the lessons learned from the front runners. From Cunard's *Lusitania* and *Mauretania* came the propulsion system of four screws driven by steam turbines; from White Star, the details of *Olympic*'s extravagant interiors and, just in time, the lessons learned from the *Titanic* disaster. There was, too, the incontestable advantage of assigning a length that would establish a world record. It was a source of great pride that the new liner's length of over 900ft would exceed *Olympic*'s by a comfortable margin.

Imperator, the first super-liner, nears completion in the fitting-out berth at the Vulcanwerke shipyard, Hamburg, in 1913. (Chronicle / Alamy Stock Photo)

Manoeuvring along the River Elbe, the newly completed *Imperator* is seen with her original, taller funnels and revealing the massive eagle adornment on her prow, which was emblazoned with the slogan '*Mein Feld ist der Welt*'. (Hapag-Lloyd)

It should be remembered that back in 1910, the Vulkan Werke shipyard had never constructed a ship of what were then such epic proportions – a ship that was to become the largest in the world when she entered service. The pressure was on. As lead yard there was no comparable builder to consult with, to make comparisons with or to refer back to, and the fact that the project had to be right first time led to financial difficulties that Vulkan encountered as the build progressed.

On 13 May 1912 *Imperator* was launched by the Kaiser, and when the hull became waterborne the port anchor was released to arrest her travel. However, the complete anchor and chain paid out, because either the chain had snapped or it had not been secured to its post in the chain locker.

During fitting out it became known that Cunard's new unnamed ship, *Aquitania*, was to be 901ft long, a restricted statistic that Cunard had been trying to keep quiet about. Although *Imperator*'s overall length would be 909ft, Ballin was keen to have his new vessel declared the 'world's largest and longest'. To this end he commissioned Berlin sculptor, Dr Bruno Kruse, to design and cast a gilt and bronze figurehead for her. It was a crowned German eagle astride a globe with the HAPAG motto '*Mein Feld Ist Die Welt*'. When mounted on the bow it brought *Imperator*'s overall length to 919ft. However, it was later removed after being damaged by a large wave during an early crossing that ripped off its wings. This effectively reinstated her length to 909ft but later some records still recorded her dimensions as 919ft.

At the time, with the *Titanic* disaster still firmly in mind, in her construction *Imperator* provided the best safety features. A double watertight skin ran around the ship from the double bottom upwards to a point above the waterline. In addition, wing coal bunkers protected the boiler rooms as it did in *Mauretania*. Transverse bulkheads formed thirteen transverse watertight compartments, whilst further longitudinal bulkheads split up the engine room compartments. Access between compartments was provided by thirty-six watertight doors, hydraulically activated and electrically controlled from the bridge employing a system developed by Stone-Shand. Doors could be closed or opened together or separately from the bridge but local control, by means of a lever, enabled individual doors to be operated when necessary. Gongs sounded at each door when it was closing. The hydraulic force was sufficient to crush any coal lumps, which was essential for bunker doors. There were also enough lifeboats for all 5,277 people on board at full capacity. When she entered service there were eighty-three lifeboats stowed aboard, two of which were motor driven, sufficient for 5,500 passengers and crew.

Ballin also wanted his new class to be speed contenders if they were to maintain the three-ship service. He broke from the tradition of steam reciprocating prime movers and opted instead for steam turbines. Four sets of AEG-Curtis-Vulcan-built steam turbines were installed, manufactured under licence from Parsons, Newcastle upon Tyne. Together they gave a total of 74,000 shp. Direct-drive turbines provided the only feasible means of propelling the ship, but they differed from the conventional passenger ship form of the time. Triple-expansion turbines were chosen, there being two LP (low-pressure) turbines connected to the outboard shafts, an IP (intermediate pressure) to the starboard centre shaft and an HP (high pressure) to the port centre. For astern manoeuvring, the HP turbines turned the centre shafts and the LP units the wings, giving an astern power output of 35,000 shp. To convey an impression of their scale, in total the turbines contained some 760,000 bronze blades. For normal operation, steam entered the HP stage, which exhausted to the IP turbine, which in turn exhausted to both LP turbines. Should maximum power have been required, steam could also be routed through the IP turbine. Steam at 235 psi was supplied by forty-six Yarrow Naval-style water-tube boilers, forced draught by Howden and originally coal-fired, to give the ship a speed of 23 knots.

Imperator was delivered on 23 May 1913 with an accommodation for 908 first-, 972 second- and 942 third-class passengers, together with 1,772 passengers in steerage, from unashamed luxury to

Imperator anchored in New York Harbour on 19 June 1913 during her maiden voyage, seen from the bow and stern quarter. She is probably waiting for the tenders to disembark immigrant passengers and their luggage for Ellis Island before proceeding to her Hudson pier. (US Library of Congress)

Left: Imperator sailing from an unidentified port, possibly Cuxhaven. (Richard de Kerbrech collection)

Below: When *Imperator*'s eagle figurehead was removed, after being damaged while crossing the Atlantic in heavy weather, it was replaced as seen here by an ornate decoration on either side of her bow featuring the Imperial crown. (Ludolf Timm)

the extremely basic. Upon her entry into service she was truly the world's largest ship at 52,117 gross registered tons and was nicknamed 'Colossus of the Atlantic'.

The advent of such a big ship as *Imperator* made it an urgent necessity to have the clearest approach to New York harbour. Tortuously, the Ambrose Channel Dredge scheme was completed after twelve years' hard work, just as she was made ready for sea. The Channel had been used for some time past and was just big enough to accommodate *Lusitania* and *Mauretania* when they came out. However, the New York port authority was not satisfied and aimed at nothing less than the completion of the full scheme, a 7-mile channel with a minimum width of 2,000ft and a depth of 40ft of water at low-water spring tides. *Imperator*'s draught was 35ft 6in.

Imperator sailed on her maiden voyage on 10 June 1913 from Cuxhaven to New York via Southampton and Cherbourg. Despite being fitted with Frahm Anti-Rolling Tanks, her inaugural voyages were fraught with stability problems and she was 'tender' in a seaway. Possibly her luxurious and voluminous public rooms built high in her structure with very high ceilings may have contributed to her having a small metacentric height. To help rectify these problems, she was returned to her builders in November 1913. Nine feet were taken off each funnel, the Grill public room was removed, and 2,000 tons of concrete was added to her permanent ballast. The improvements were due to be put to the test on her first transatlantic voyage after the refit, which departed Cuxhaven on 11 March 1914. On 1 August 1914 she was scheduled to sail from Hamburg to New York, but this voyage was cancelled and she was laid up at Hamburg for the duration of the First World War.

Fast-forward to 5 May 1919 when the US Navy Department commissioned the seized *Imperator*, which had survived the war, as USS *Imperator* (ID-4080) to operate as a troop transport at Brest in western France. By then she had become the second largest ship in the world. Found in good condition, after weeks of dredging had pulled her loose from the Elbe River's mud, she was steamed to Brest from Cuxhaven on 27 April for the formal transfer. Upon her arrival at Brest on 5 May she was placed in commission with the US Navy, joining her younger and larger sister USS *Leviathan* (formerly *Vaterland*). With both ships under the US Ensign, not only were they the two largest passenger ships in the world but also the two largest warships in the US Navy. Their mission was to repatriate all US troops from the European theatre of war. Adaptation for dormitory accommodation of troops with provision of extra toilet facilities followed, besides the taking on of stores and the filling of her coal bunkers, after which USS *Imperator* embarked some 1,500 army officers, 300 enlisted men and 500 nurses. On 15 May, she sailed in company with USS *Leviathan* for Hoboken, New Jersey. Although not declared as such, both ships' companies and troops viewed the crossing as a race between them to relieve the shipboard boredom arising from being stood-down troops.

Still sporting her bow decoration, *Imperator* is seen while serving as a troopship under the Stars and Stripes, arriving at Pier 4 Hoboken, New Jersey, from Brest on 26 May 1919, at the end of her first voyage since 1914. Here she is carrying 3,190 American servicemen. (US National Archives 111-SC-53847)

A later view shows her commissioned as USS *Imperator*. She is being assisted to a berth at Hoboken, New Jersey, in mid-1919 after arriving from Brest. (US Naval Historical & Heritage Command NH105815)

Operating with the Cruiser & Transport Force from 3 June to 10 August 1919 she made three round trips from New York to Brest, returning over 25,000 troops, nurses and civilians to the United States. While en route to New York on 17 June, USS *Imperator* assisted the French cruiser *Jeanne d'Arc*, which had broken down in mid-Atlantic. President Moreira of Brazil had been travelling on board *Jeanne d'Arc* and *Imperator* conveyed him and his party onwards to the United States.

On 17 August 1919 *Imperator* was scheduled to sail from Hoboken on her last trooping voyage and then to be laid up, after which she was to be returned to Germany. She was handed over to the British Ministry of Shipping on 24 November 1919. There was much controversy over what to do with her but Britain was pressing a claim for her as a replacement for *Lusitania*. By January 1920, the issue was settled in Britain's favour. Under the Treaty of Versailles the United States would retain *Leviathan* while *Imperator* and the still incomplete *Bismarck*, the third ship of the Hamburg Amerika trio, would be ceded to the UK. *Imperator* arrived at Southampton on Sunday, 10 December 1919, before proceeding to Liverpool for drydocking on 6 January for what was intended to be a quick overhaul. However, upon inspection, issues with her rudder and propeller erosion were discovered. The remedial work necessitated her retention in Liverpool, causing the cancellation of her inaugural Atlantic crossing for Cunard planned for 10 January 1920. Due to the extent of the work that had to be carried out, *Imperator* remained at Liverpool for around seven weeks and the Cunard company dinner of January 1920 was held in the dining room aboard her.

The following month she was chartered by Cunard from the Shipping Controller and made her first voyage for Cunard from Liverpool for New York on 21 February. Later, in June that year, she was switched to the Southampton–New York run.

In a combined acquisition with White Star, *Imperator* was awarded to the Cunard Line. She served under her original name until 16 April 1921 when she was renamed *Berengaria*. She is seen with tugs at Southampton under her German name (Authors' collection)

During her troopship role, the crew had included half a dozen German guarantee engineers but upon Cunard's acquisition they were relieved of their duties. It then transpired that her tender stability had again manifested itself. Near the end of a second voyage while two days out of Cherbourg, she encountered heavy weather whereby seawater entered the ash ejectors and clogged up the pumps. Some of the coaling port seals leaked, causing water in the bilge to rise rapidly, resulting in the ship taking on a list.

In February 1921 she and the incomplete *Bismarck* were bought jointly by the Cunard and White Star Lines, both companies having a half share in each ship but being in complete control of only one of them. *Imperator* was renamed *Berengaria* and sailed from Southampton on 16 April 1921, inaugurating a vital but austere service. However, in September 1921, with *Aquitania* reconditioned and *Mauretania* just ready after conversion to burn oil, *Berengaria* was sent to Armstrong, Whitworth & Co. of Newcastle upon Tyne, where she underwent a much-needed refit.

A stern view of *Berengaria* at the time she was refitted on the River Tyne. The decorative scrollwork around her stern, fitted to her as *Imperator*, was retained. (Tyne & Wear Museums)

Berengaria was sent to Tyneside in October 1921 for renovation and conversion to oil fuel by Armstrong Whitworth. In order for the world's third largest ship to navigate the River Tyne it required eight tugs: four towing, three at the stern and one in reserve. (Tyne & Wear Museums)

The newly named Cunarder arrived on the Tyne on 31 October 1921 for a six-month refit and conversion. She was given a thorough refit and converted for oil fuel. In this her side bunkers were changed over for the carriage of oil, these side tanks extending for a length of 300ft. Altogether there were twenty-four storage tanks along with two settling tanks per boiler room, each of these holding between 150 and 200 tons. The total oil bunkers carried amounted to 6,500 tons, from which her daily consumption was around 700 tons per day. All the boilers were converted to oil burning, involving much retubing.

Slight external changes were made by shifting the aftermost three lifeboats stowed on each side under the promenade deck up on to the boat deck directly above. The boats on the short poop deck aft were removed. In all, about 1,000 tons of extra steelwork was put

Berengaria in the River Hudson, making an eastbound departure. (Authors' collection)

Outward-bound for the United States, *Berengaria* passes the Bramble Bank in The Solent. In the background is the inward-bound Red Star liner *Belgenland*. (Smith Archive / Alamy Stock Photo)

into the ship. The accommodation layout of steerage and third class was altered drastically, whereby the steerage spaces were stripped out altogether. All the marble bath tubs in first class were replaced by the galvanised iron type and an extra 1,000 tons of pig iron ballast was added. She emerged with refurbished accommodation for 972 first-, 630 second-, 515 tourist- and 606 third-class passengers.

In the meantime, at Southampton, work had commenced on hewing out a 'vee' notch at the land end of the No. 6 Trafalgar Drydock in order that it could accommodate *Berengaria*'s length during future drydocking.

Berengaria left the Tyne on 26 April 1922 and in early May she entered the Cunard 'Express' service from Southampton to New York along with *Mauretania* and *Aquitania*. Perhaps 'Express service' was a bit of a misnomer for, along with the ships *Carmania* and *Caronia*, all five had been built pre-war and in effect were a skeleton fleet that was probably incapable of meeting the demands of a fast express Atlantic service as compared to the past. They could not match the competitive strength of other liners in this specialised business. That said, the same year *Berengaria* made a run from Cherbourg to Ambrose at 23.27 knots and an eastbound passage of 23.79 knots. One of her masters, Captain Harry Grattidge, described her as 'principally a gleaming and bejewelled ferryboat for the rich and titled', and in the roaring twenties she attracted the sobriquet of 'the Millionaires' Liner', remaining the flagship of the Cunard Line until 1934.

The *Manchester Guardian* reported on 25 September 1925 that a security alert at sea had been triggered when the Cunard offices in New York received a message stating that there was a bomb on *Berengaria*, which had 1,600 people on board. The ship was then 1,200 miles out from New York en route for Southampton. The ship was searched systematically, although the passengers and most of the crew were not informed as to the reason. A fire drill was held just before the scheduled time of detonation, so that passengers could be placed close to their lifeboat stations without panicking them. The bomb threat failed to materialise though and no one claimed responsibility.

Berengaria at speed, a view from the 1920s. Southampton taxi drivers nicknamed her the 'Dead 'n' bury her' but she was also referred to as 'the Happy Ship'. (Cunard Archives, University of Liverpool)

Berengaria was used during April 1928 as the reception point in mid-Atlantic for the first ever shore-to-ship transmission of television images, making an important contribution to developments in telecommunications. The pictures were transmitted from John Logie Baird's television studio in London via a short-wave station in Surrey and successfully received aboard the ship by the Chief Wireless Operator, S.W. Brown.

Early in 1929 a New York stockbroker, Michael J. Meehan, successfully negotiated with Cunard to try out a floating exchange aboard the liner. Two dealers and an extra radio operator were recruited and service arrangements were made with radio stations at Cape Cod and New Jersey. All went well for a few months but on 29 October 1929, while the ship was bound for New York, the Wall Street market collapsed. Thousands were ruined, including many of *Berengaria*'s passengers. It signified the end of the floating exchange. During the Depression that followed, *Berengaria* was diverted to short $50 cruises from New York. Taking place during the Prohibition era, they became known as 'booze cruises' and the ship received a new nickname: 'bargain area'.

The tug tender *Romsey* assists *Berengaria* to her Southampton berth. (*Southern Evening Echo*)

In February 1932 the crime novelist and writer Edgar Wallace died while in the United States working on a film script. Later his body was brought back to the UK aboard *Berengaria*, which sailed up Southampton Water to the docks with her flags at half mast. Two years later, in February 1934, Cunard and White Star merged to form the Cunard White Star Line, and following this *Berengaria* received an extensive refit. That year Captain Edgar Britten was appointed as her commander.

In May 1936 *Queen Mary* entered service, essentially offering a four-ship service. On 20 October that year *Berengaria* had a fire outbreak in several cabins during an overhaul in Southampton, and when this damage had been rectified she relieved *Aquitania* on 18 November, running in consort with the new '*Queen*'.

By then she was in her run-down period but, even so, during 1937 *Berengaria* made fifteen westbound crossings carrying a total of 11,211 passengers and fifteen eastbound crossings with a total of 10,555 passengers.

While in New York on 3 March 1938, *Berengaria* was damaged by a fire in her main (cabin-class) lounge and other passenger accommodation areas. The US authorities were not happy, regarding her to be a fire risk, and as a consequence she returned to the UK without passengers and was laid up in Southampton's Western Docks.

She was put up for sale in October 1938 and on 7 November was bought by Sir John Jarvis to be taken to Jarrow for scrapping as a means of helping alleviate the unemployment in the area. By 1939 the ship was dismantled down to her double bottom, forming a sort of raft. In 1946, following the Second World War, what remained was cut in two and these sections were towed to Rosyth to complete the breaking up.

Berengaria laid up at 108 berth, Southampton, in Easter 1938, prior to her departure to the breakers yard on the River Tyne. She was then the oldest of the former HAPAG trio and the last to be withdrawn from service. (Southampton City Museums)

THE SECOND MASSIVE SUPER-LINER: *VATERLAND*

Blohm & Voss had secured the order for the second of Ballin's trio, which was to become *Vaterland*. They were possibly disgruntled at having not been awarded the contract for *Imperator*, but they could benefit from experience gained from the earlier ship with regard to major design points. Dr Ernst Foerster led the design of the second ship. As a result, with two extra frames introduced into its construction, the mighty hull had an extra 2ft on the beam and was 25ft longer, resulting in the ship having an overall length of 949ft 9in. It may be that serious hull fractures in later years could have been the result of this modification.

Another innovation was that the boiler uptakes from the forty-six boilers were divided to port and starboard and re-joined just below the two forward funnels. This left a large central space that gave the designers the opportunity of laying out the magnificent public rooms on the upper decks with a tremendous vista forward and aft down through the centre of the ship. This same feature was incorporated into the third ship of the trio, *Bismarck*. On each side of *Vaterland*'s stem were shields carrying the arms of Germany.

Vaterland, Yard No. 212, was launched at the Blohm & Voss shipyard on 3 April 1913 and upon her completion on 29 April 1914 she became the largest ship in the world at 54,282 gross tons. During her trials she was reputed to have reached runs of 25.8 knots.

On 14 May she departed Hamburg on her maiden voyage to New York via Southampton and Cherbourg. She joined *Imperator* and *Kaiserine Auguste Victoria* on the HAPAG service but it was a short-lived career, completing only three round voyages. While on her fourth westward bound crossing, one of *Vaterland*'s four main turbines failed, causing her Master, Captain Hans Ruser, to shut it down and run on three screws. At the same time, all four of her astern turbines malfunctioned, disabling her low-speed manoeuvrability, and she had to rely on the New York tugs to dock at Hoboken on 30 July, just before the outbreak of the First World War.

Launch of Hamburg Amerika Line's second giant, *Vaterland*, at Blohm & Voss on 3 April 1913. Her underwater hull form, with a block co-efficient of 0.625, was fine by future liner design standards. (Frank O. Braynard)

Vaterland's bow, seen in drydock at Hamburg. (Authors' collection)

Two views of *Vaterland* during her brief commercial career: the first, above, taken in The Solent on 15 May 1914 while making her maiden call at Southampton (Beken of Cowes); the second, below, showing her arrival at New York on 29 July 1914 (US Naval Historical & Heritage Command NH103156)

British warships were already off the United States coast ready to seize German liners trying to make it back to Germany. *Vaterland* was too big and valuable to risk, so she remained at New York for three years along with other laid-up German ships, although her crew enjoyed a certain amount of freedom through the period in which the United States remained neutral. As it became evident that the US would enter the war against Germany, it was thought that some of her crew might attempt some form of sabotage in anticipation of seizure. After almost three years of idleness, the US government did seize *Vaterland* on 4 April 1917. It was decided she was most suitable for use as a troopship and she passed to the US Navy's control during June that year.

The naval inspectors found the liner to be structurally sound, but the mechanical problems that had plagued the ship on her last pre-war crossing had not disappeared during her idleness. Although it was believed *Vaterland*'s propulsion machinery had not been purposely sabotaged by her crew, as was initially suspected, improper maintenance and operation during her first year of service, and neglect during her internment, had taken their toll. Her turbines were in bad shape, as were her boilers, the auxiliary machinery and piping. Indeed, one astern turbine was found inoperable with its blading stripped off and the casing cracked. A US naval officer and a marine engineer of the day described the engine rooms as a 'nightmare'. In addition, the ship's hull was encrusted with several years' growth of barnacles and the silt build-up that anchored her to the bottom of her berth. It is worthy of mention that the ship's power plant was not only extremely intricate but did not contain universal component parts and there were no replacements readily available in the US. Any that were required had to be designed and manufactured or a particular item of equipment completely replaced by a US equivalent. The power plant was so unusual and complex that no one in the US was really familiar with its operation. To add insult to injury, none of the ship's former German engine room staff was prepared to co-operate by providing instructions. The ship's drawings were in the hands of Blohm & Voss, who refused to sell them for less than £200,000. The

Commandeered as USS *Leviathan*, the former *Vaterland* is seen in the Gladstone drydock, Liverpool, where she received her coat of dazzle paint. The photograph dates from mid-1917. (US Naval Historical & Heritage Command NH105518)

Sporting her dazzle livery, USS *Leviathan* alongside at Hoboken in 1918. The dazzle design was created by Norman Wilkinson's team in the Royal Academy workshops, London. (US National Archives)

After her dazzle livery had been replaced by an overall coat of grey paint, USS *Leviathan* arrives at New York on 21 December 1918 while ferrying US troops from France. (US National Archives 165-WW-138A) The following year, she made another New York arrival on 7 May. That year, *Leviathan* transported 9,470 troops on one voyage: the record number for the First World War. (US National Archives 111-SC-48217)

The United States retained *Leviathan* and she was sent to the Newport News Shipbuilding and Drydock yard in Virginia for rebuilding as a passenger liner. Here she can be seen during the renovation work as the yard employees head ashore at the end of their shift. Note the post-war change to her lifeboat numbers and their disposition. (Frank O. Braynard)

Leviathan arriving at New York attended by numerous tugs, long before the days when even bigger liners berthed themselves unassisted, using their bow thrusters. (Frank O. Braynard)

The Boston Navy Yard drydock No. 3 was the facility used for refits of the former Ballin trio of super-liners. Here *Leviathan* is secured in the dock. (US Navy)

US authorities refused to consider any such sum. Fortuitously, a Scots engineer representing the Parsons Marine Turbine Co., under whose patent the turbines had been originally built, was a resident engineer in the US and he was able to offer his expert consultancy.

Urgency was required to refit her as a troopship. The first step was to repair and recondition her power plant. Working without detailed engineering drawings, navy repair teams were able to bring the ship's boilers and turbines back on stream. The replacement of damaged piping, where several water pipes behind the panelling had in fact been deliberately cut and crimped, sabotaged by the German crew, to manifest slow leakage over a long period, and also the re-routing of much electrical cabling, was achieved with surprising speed. All her luxurious furnishings and fittings were removed ashore and cohorts of pipefitters, carpenters and plumbers installed thousands of standee bunks. The cost of surveying the ship and draughting as-fitted drawings was £50,000.

As the on board conversion was under way, US Navy salvage specialists were examining ways to extricate the massive ship from her silted-up berth. The Army Transport Service brought in dredgers, and after four days non-stop work under direction from divers, she was freed. Her conversion work was completed on 25 July 1917. The US repairers had never dealt with a passenger ship of such dimensions before and *Vaterland* needed to be drydocked with some urgency. Without such a facility available Stateside to accommodate her, she was scheduled to have the cleaning of her hull and propellers done in Liverpool's Gladstone Graving Dock after her first eastbound crossing.

Officially commissioned into the US Navy, she was renamed USS *Leviathan* on 6 September 1917. She underwent a shakedown cruise to Cuba and attained a speed of 23 knots, then returned to Hoboken to have gear, 6in guns and machine guns fitted.

Ready on 15 December 1917, she departed Hoboken with 7,250 troops aboard bound for Liverpool, where they disembarked. *Leviathan*'s draught of 39ft 6in was a limiting factor and she could only just clear the drydock sill on the high-water spring tide on 15 January 1918. While the hull and propellers cleaning was under way in the dock, sixty men worked arduously to apply her dazzle paint scheme. A triangular appendage was also fitted to the foot of her bow. This was installed to permit the attachment of paravane wires as a defence against tethered mines. After twenty-six days she left the drydock on the next spring tide and so began her career as a troopship.

Though successful, *Leviathan*'s first two round-trip crossings highlighted the problem of her deep draught restricting her entry into Liverpool to only during the high spring tides of new or full moon. This limited her trooping activities to just one round-trip voyage every two lunar months. Naturally, this restricted the number of troops the ship could contribute to the war effort and enforced idleness upon the Allies' largest and most capable asset.

It was suggested in April 1918 that *Leviathan*'s eastern terminal should be Brest in Brittany rather than Liverpool, as the deep-water French port was not affected by the tides. This enabled *Leviathan* to speed up her transatlantic shuttle and significantly increase the number of US troops dispatched to the battlefields. The other benefit of sailing directly to Brest was it eliminated the logistical chore of transporting offloaded troops in the UK onwards overland and by ferry to France. In addition, the need to coal ship in Liverpool could be eliminated by shipping this to Brest. In such a manner the disembarkation of troops and bunkering were carried out simultaneously, thus shortening *Leviathan*'s turnaround time. This decreased the average turnaround time from fifty-four to twenty-six days. Coupled with the increase of her troop capacity to 10,500, this resulted in an annual increase of 30,000 troops carried to France by the ship.

The Great Spanish Influenza pandemic of 1918–19 took its toll on those aboard *Leviathan*. The disease had spread worldwide and was rampant in the staging areas where US troops were marshalled prior to being sent overseas. *Leviathan* left New York on her ninth voyage on 29 September 1918 with 9,320 troops and a crew of nearly 2,500. The flu had apparently been brought aboard because on the second day at sea the ship's small hospital was full. On 2 October, 615 additional bunks had been occupied by the sick, and the first death occurred on that day. By the time she reached Brest, on 7 October, ninety-three more soldiers had died. A further fifteen died after being landed and the ship's medical officer estimated that more than 3,000 of the 11,820 men aboard were seriously ill. Notwithstanding this, the epidemic did not interfere with *Leviathan*'s trooping duties.

Leviathan outbound in The Solent on 17 July 1923 following her maiden call at Southampton. (Beken of Cowes)

By the time she had completed her military service on 8 September 1919, *Leviathan* had transported 192,753 people to and from Europe during her nineteen transatlantic crossings. It is believed that on one voyage she carried a total of 13,558 persons! Upon her final return she was decommissioned by the US Navy on 29 October 1919 and handed over to the United States Shipping Board. For the next two and a half years she was laid up in Hoboken pending a decision as to her future. She was ultimately transferred in February 1922 to United States Lines management on behalf of the US Shipping Board and a contract was awarded to the Newport News Shipbuilding & Dry Dock Company, Virginia, for her complete reconditioning after she was moved there in April that year.

Leviathan seen from above as she arrives off Manhattan. (United States Lines)

Leviathan prepares to sail from Southampton's Ocean Dock on 29 August 1923, bound for New York with 2,587 passengers aboard. This was a record carrying on the westbound run for the previous three years. (United States Lines)

This proved to be a massive undertaking. Her accommodation was almost completely renewed, providing for 876 first-, 548 second-, 944 third- and 936 fourth-class passengers and a crew of 1,150, of whom 693 were in the catering department. The ship was practically rewired throughout, additional strengthening was given to the hull and improved fire protection was installed. In the engine room her machinery was completely overhauled and her boilers were converted from coal to oil burning, with bunkers for 9,500 tons. This effectively reduced her engine room staff by 225 men. The refit and overhaul cost around $8 million and resulted in *Leviathan* becoming once more an exceedingly fine ship. The task was completed on 19 June 1923 and at that time her gross tonnage was remeasured under US rules at 59,956 gross tons, permitting her to claim she was the largest ship in the world. During trials she achieved a speed of 27.48 knots.

Over in the UK, British seamen had been enduring annual wage cuts since the First World War, cuts that the British shipping lines deemed necessary to stay afloat as the nation struggled to recover from the crippling cost of the war. While *Aquitania* was in New York on the same day as *Leviathan* was preparing to sail, a party of her stewards 'jumped ship' and went to sign on *Leviathan*, doubling their wages. When *Leviathan* arrived in Southampton, though, the Borough Police there were waiting ready to arrest the defaulters for breaking their articles.

Another fine view of *Leviathan*, a photograph taken of her at New York. (R. Hildebrand)

Leviathan had resumed transatlantic service on 4 July 1923, running in service with *George Washington*, *America* and *Republic*, but her greater speed over her consorts resulted in an imbalanced service. It is believed that on her first voyage she showed a net loss of $100,000, a situation that continued on each subsequent voyage. It was the time of Prohibition in the US (1918–33) and she was obliged to run as a 'dry' ship; moreover, as the 'Roaring Twenties' came to a close through the Great Depression, this had a negative effect on her passenger numbers.

Up until 1927, things had been progressively looking up for *Leviathan* and in that year she made fourteen voyages, or twenty-eight crossings, and carried 40,534 passengers, an average of 1,448 per crossing. On a single westbound crossing that August she carried 2,733 passengers.

In 1931 her gross tonnage was remeasured at 48,596, mainly by a manipulation of the US classification regulations. Under her original measurements it had cost the owners over $560 thousand (*c.* £100 thousand) in increased drydock and harbour dues. Not only this, but she had been dogged by structural damage that year in which her strength deck plating fractured at the corner of the deck opening and under the midship expansion joint. She was laid up as unprofitable in 1932 and 1933, and made only five voyages in 1934, similarly in 1935. In June 1936 she resumed service after reconditioning at a cost of £30,000, it being estimated that some £12 million had been spent on her by her American owners since the First World War. Later, on 14 September that year, she was withdrawn from service after five round trips, having made heavy losses, and was laid up at Hoboken.

During July 1935 a proposal was put forward whereby *Leviathan*, along with *George Washington* and *America*, should be used as training depots for US Merchant Marine officers and crews. However, this did not materialise and in December 1937 she was sold to British shipbreakers Metal Industries Ltd of Rosyth and Thos. W. Ward of Sheffield.

Leviathan under way at sea. (US Naval Historical & Heritage Command NH103130)

After three years' lay-up and neglect, she left Pier 4 in Hoboken and departed New York on 26 January 1938 for Rosyth, flying the Red Ensign. The once proud liner was eased out by tugs but this time no jazz band played on her promenade deck, no passengers lined her rails and on the pier there were only a handful of stevedores to cast her off. In addition to her rust-streaked white superstructure, her funnels and masts had been cut down to enable her to pass under the Forth Bridge. On board were 150 American and British crew members under the command of Captain John W. Binks, who had formerly commanded *Olympic*.

Leviathan preparing to leave her Hoboken Pier 4 on 25 January 1938 on her final Atlantic crossing, bound for Scotland for breaking up. Her funnels and masts have been cut down to permit clearance under the Forth Bridge en route to Rosyth. As the removal of these structures had taken place while alongside at Hoboken, it raises the question as to how they were lifted off the ship in the absence of cranes. (Bill Miller and Richard de Kerbrech collection)

On the way across the Atlantic, boiler tubes burst or leaked daily; after starting out on thirty-six boilers, she ended up on only about twenty-four. Speed on occasions dropped to zero, two oil fires occurred in the boiler rooms and fractures opened up again in her strength deck, reaching alarming proportions. *Leviathan* sprang a leak in a compartment right aft that proved very difficult to pump out. Coupled with this, the transferring of ballast water and oil fuel, necessary to ease the stress on the strength deck, resulted in a change of trim that made the final stage of the voyage very precarious. Crew conditions during the crossing were spartan, with filthy accommodation, a lack of domestic hot water, and food shortages. *Leviathan* arrived in the Firth of Forth on 3 February and had to remain anchored below the Forth Bridge for ten days before proceeding up to Rosyth on 14 February to her final resting berth for demolition.

For all her reliable and regular yeomanry service in her years as a troopship, *Leviathan* had a most chequered post-war career. She was unable to fulfil her full potential by providing a clockwork service working in consort with much slower fleet mates; she was plagued by high manning costs and dogged by recurring structural fatigue problems. And the Prohibition era defined her period of service, rendering her a 'dry' ship.

The same year, 1938, Mr George Gaede wrote of *Leviathan*: 'The competitive life of a ship is twenty years at most. When the [US] Government placed the vessel in commercial service, the best years of her life had already passed, and the necessary reconditioning placed her even farther behind in the parade on the Atlantic lanes.' It was a scathing comment stated in hindsight.

BISMARCK: THE LAST OF THE TRIO IS COMPLETED – BUT NOT ACCORDING TO PLAN

Bismarck was the third of Hamburg Amerika Line's giants and as such was the last of an era of mammoth liners planned before the First World War. The late Chancellor Bismarck's granddaughter, Countess Hannah von Bismarck, launched her on 20 June 1914 at Blohm & Voss' Hamburg Yard (Yard No. 214). At the climax of the ceremony, the Countess experienced some difficulty in breaking the champagne bottle on *Bismarck*'s hull so Kaiser Wilhelm II, who was also present for the naming ceremony, stepped forward and broke it for her.

Within weeks of the launch, the outbreak of war interfered with *Bismarck*'s progress as all shipyard labour was diverted to more urgent war effort. Thus, she languished in an incomplete state for the duration of the conflict. In this situation, most of her brass and copper fittings were cannibalised for use elsewhere.

Some five years later on 28 June 1919, under the Treaty of Versailles, the British Reparations Committee awarded her and *Imperator* to Great Britain to replace the sunken *Britannic* and *Lusitania*. To this end, the White Star Line, aided by Harland & Wolff technical staff, tactfully set about the task of supervising the completion of *Bismarck* to their own specifications at the Blohm & Voss shipyard in Hamburg. On 5 October 1920, her completion was delayed considerably by fire. Sabotage by German shipyard workers was at first suspected as they were understandably reluctant to part with her. Notwithstanding this, the 1,200-strong shipyard workforce turned to in order to complete her construction. Her dimensions surpassed those of her two earlier sisters with a gross tonnage of 56,551, an overall length of 955ft 10in and a 100ft beam, making her the largest ship in the world, notwithstanding *Leviathan*'s dubious claim. The ship's depth from her keel to the boat deck was 101ft. Initially the *Bismarck* was bought under a joint package deal by Cunard and White Star Lines, but as Cunard were already operating *Imperator*, now named *Berengaria*, *Bismarck* became a White Star ship during February 1921.

Through the delay in her construction, she was able to be completed as an oil-fired ship, unlike her two predecessors. On 28 March 1922, *Bismarck* was completed and White Star crew and officials under Captain Bertram Hayes arrived at Hamburg to take her around to Liverpool for trials. They found the liner alongside the dock painted in Hamburg Amerika Line livery and boldly displaying the name *Bismarck*. This had been allowed to happen to assuage German feelings. On board, the Master found his cabin being used for the storage of ship's fittings. Furthermore, they found that royal suites had been built into the ship to carry the entire German Imperial family on a round-the-world victory cruise!

Members of the German public and the shipyard workforce were reported to have lined the dockside and banks of the River Elbe in silence when *Bismarck* departed Hamburg for Liverpool. According to Captain Sir Bertram Hayes, upon *Bismarck* clearing Cuxhaven, he had the German name painted over on the bows and stern and the name *Majestic* painted in its place.

On 1 April, *Majestic* commenced a series of trials in the Irish Sea that lasted over a period of ten days. Upon their completion she was officially named *Majestic* on 12 April. From Liverpool she then sailed around to Southampton to be loaded with stores and have her White Star livery painted prior to her maiden voyage, marketed as the world's largest liner. She sailed on her maiden voyage from Southampton to New York via Cherbourg on 10 May 1922, under the command of White Star Commodore Hayes, who had been promoted in May 1922.

Majestic had nine decks and her hull was subdivided transversely by watertight bulkheads into fourteen compartments, the doors being controlled from the bridge, while her hull consisted of cellular construction that gave great strength to the vessel. As with *Vaterland*, the boiler uptakes did not lead directly into the two forward funnels through the central part of the ship. Instead, they were divided and led up the sides of the ship to rejoin just below the funnel base, above the level of the boat deck. The after funnel was used solely for ventilation purposes.

This configuration gave a vast, uninterrupted area of open space on the promenade deck which was also used to great advantage in her magnificently decorated public rooms. Her interior space was said to be equivalent to 400 eight-room houses. Of her decks, the five lowest ran the full length of the ship and the watertight bulkheads extended up through them. The first-class dining room, which was also the central hall, was situated on F deck and was the largest ever built in a ship. The central part had an area of 2,300 sq ft, was 31ft high and was topped by a lofty, frescoed dome supported on slim ionic pillars. The total area of the room, which had a seating capacity for 678 passengers, was 11,350 sq ft. The dining room was surrounded by tall windows that helped illuminate its interior. Above the fifth deck were the four decks that contained the cabins and public rooms like the first-class lounge, the à la carte restaurant and the palm court. One of her luxury amenities was the swimming bath in the Pompeian style of ancient Rome. It was furbished with rich marbles and brilliant red mosaics and had an area of 820 sq ft, with a depth ranging from 3 to 9ft.

The third and largest of the great trio of super-liners commissioned by Albert Ballin for the Hamburg Amerika Line, *Bismarck*, on the slipway at Blohm & Voss, Hamburg, on launch day, 20 June 1914. The dignitaries attending the occasion can be seen on the right, mounting the steps to the covered podium from where the ceremony was performed. (Blohm & Voss)

Like *Imperator* and *Vaterland*, *Majestic* was powered by quadruple screws driven directly by Parsons steam turbines, with an astern turbine on each shaft. As with the two previous ships of the Ballin trio, the four turbines normally worked in triple combination. The steam generation flow and propulsion system were identical except that steam at 260 psi was supplied by forty-eight Yarrow-Normand water-tube boilers with a total heating area of 220,000 sq ft. The boilers were fired with oil fuel on the J. Samuel White system, that is with five burners fitted to each boiler. As before, the turbines developed a total of around 80,000 shp.

The bunker compartments were in tanks on each side of the boiler room, a cross bunker forward and in double bottom tanks, together providing capacity for 9,000 tons of fuel oil. The fuel consumption on each Atlantic crossing was around 5,700 tons. Her screw propellers were four-bladed, of 16ft 5in diameter and rotated at 200 rev/min. This power plant gave *Majestic* a normal speed of about 25 knots, which was surpassed only by Cunard's *Mauretania*. She was also fitted with Frahm anti-rolling tanks in an attempt to dampen any roll in a rough seaway.

Accommodation was provided for 700 first-class passengers, located midships and forward. The second-class passengers, of whom there were 545, were accommodated aft of first class in two-, three- and four-berth cabins. The 850 third-class passengers were on four decks aft of second class.

During the first week of August 1922, *Majestic* arrived back in Southampton in the afternoon to disembark her passengers before she sailed across to Cowes the next day and anchored in Cowes Roads alongside the Royal Yacht *Victoria & Albert*. Here King George V and Queen Mary, with Prince George and the Duke of Connaught, inspected her during their visit to the Cowes Week Regatta. At the King's request, the Royal Standard was hoisted at the mainmast and this was regarded as an expression of appreciation by the King for the services of the Merchant Marine during the recent war.

Now at last White Star's 'Big Three' express service was under way, *Majestic* along with *Olympic* and *Homeric*. Sailing on Wednesdays, usually at noon, with arrival and departure from Cherbourg the same day, the ship was usually clear of the Lizard by dawn the following morning. Quay to quay, Southampton to New York, the transatlantic voyage took roughly six days with four days in port after arriving in New York on Tuesdays. The ship was then

The launch of *Bismarck*. This auspicious occasion, heralding HAPAG's three-ship express service, was soon overshadowed by the outbreak of the First World War. (Blohm & Voss)

Above left: After the end of the war, *Bismarck* was finally completed following years of inactivity. Here she is at anchor, probably at Cuxhaven, awaiting her fate. She was destined to be transferred to Great Britain to become White Star's *Majestic*. (Hapag-Lloyd)

Above right: Assisted by tugs in Southampton Water, *Majestic*, ex *Bismarck*, is shown on the occasion of her first entry into the port after leaving Germany in April 1922. (Authors' collection)

Left: Majestic seen arriving at her berth in Southampton on the same day. (PA Images / Alamy Stock Photo)

Making a Southampton departure, *Majestic* leaves the docks bound for New York. (Authors' collection)

Passing Cowes, Isle of Wight, *Majestic* outbound in the 1920s. (Kirk)

re-stored and turned around on the following Saturday, the round trip taking sixteen days.

Later, on 23 December 1922, while docking during a strong gale at Southampton, *Majestic* collided with her sister, *Berengaria*, which had its starboard quarter railings wrenched away. The following year, during September 1923, *Majestic* made her fastest eastbound Atlantic crossing between Ambrose Light vessel and Cherbourg with a crossing time of five days, five hours and twenty-one minutes at an average speed of 24.76 knots. On another crossing she departed New York on 28 June 1924 with 480 first-, 736 second- and 1,409 third-class passengers: a total of 2,625 and the most ever carried by White Star in a single crossing. This was a good sign, especially when one considers that White Star's 'Big Three' were in competition with Cunard's express trio, *Mauretania*, *Aquitania* and *Berengaria*.

In early 1924, during a routine inspection when removing a ceiling, a deck plating fracture on the port side of *Majestic*'s 'C' deck was discovered. It was located forward of the uptakes of the middle funnel in the first-class foyer. The 5/8 in-thick deck plating between the uptakes had cracked. Repairs were not carried out immediately as it was felt to be attributed to a design fault that Blohm & Voss had built into the split uptakes to the funnels, and it had occurred in light plating. The forward end of these uptakes was in line with a pair of lifts and because the corners of the lift shafts or the uptakes had not been rounded (also known as radiused), this had been sufficient for high-stress concentrations to arise, which propagated the crack. By December that year, in a rough crossing, the crack had spread rapidly enough to raise serious concern. *Majestic*'s sister, *Leviathan*, also suffered this defect.

In the August 1925 edition of *Marine Engineering* journal, it was revealed that the crack had spread and had travelled down the port side of *Majestic*. Apparently, during a westbound crossing and while rolling heavily to starboard, it had added greatly to the strain on the port shear strake. The crack began amidships, working its way outboard from both funnel uptakes, spreading to the butt straps and then propagating rivet by rivet until finally C deck was completely cracked on both sides, thus totally placing the strain on all the light

In the absence of a suitable docking capability in Britain, *Majestic*, like *Leviathan* shown earlier, used the US Navy's Boston No. 3 drydock as seen here, with the first mooring lines in place. Her first docking at the facility was on 16 November 1922. (US Navy)

inboard plating between the funnel uptakes. At that point, the crack on the port side spread its way down to reach the top of the first 16in-diameter porthole it encountered. Before the article on the crack appeared in this journal, *Majestic* was already undergoing her semi-annual overhaul during July 1925. While undergoing this, her deck plating on both sides was doubled to increase the strength, which arrested the crack. However, at the same time this occurrence brought the strength of her midships structure into question, requiring close surveillance at all times. That notwithstanding, *Majestic* was ready in time for her scheduled sailing on 12 August.

In May 1925, *Majestic* made another fast passage time of five days, averaging a speed of 25 knots and consistently reaching 27 knots for five hours! Before the advent of *Queen Mary* in 1936, *Majestic* was fast becoming a popular ship and as such became known as the 'Queen of the Western Ocean', while among seafarers she was known as the 'Magic Stick'.

At first drydocking became a problem for *Majestic* since it had not been possible for her to enter Southampton's Trafalgar Drydock because she was 37ft longer than the maximum length permissible. This was even though it had been modified for *Berengaria*. Instead, when required, she used the United States Navy drydock at the Boston Navy Yard facility, commencing on 16 November 1922 for routine maintenance and to have her name mounted properly. This was followed seventeen months later when she left Southampton on 9 April 1924 and arrived at Boston on 15 April. A year later, she was able to take advantage of the Southern Railway's new 60,000-ton floating dock, which was moored off Southampton's Town Quay. She entered this for the first time for scheduled maintenance and overhaul on 26 March 1925.

The following year, on 5 June 1926, she sailed with her accommodation regraded as first, second, tourist and third class. By January 1928 she had undergone a major refit and overhaul, in the course of which she was reboilered by the Harland & Wolff works in Southampton Docks. During this overhaul the black tops of her funnels were extended further down the sides and the black hull

Later, but before the construction of the King George V Graving Dock, liners were lifted out of the water in Southampton's massive floating dock. Here *Majestic* can be seen high and dry. In the background, *Olympic* can be seen at the Western Docks. (Frank O. Braynard)

Majestic passing the RAF station at Calshot Spit during preparations for the annual Schneider Trophy seaplane race. The photograph dates from September 1929. (Science Museum)

Two views of *Majestic* approaching and then entering the new King George V drydock, the first ocean liner to enter the dock on its completion in 1934. (Both Authors' collection)

paint was extended up one deck with just her spirket plate forward painted white. Following this refurbishment, which was carried out in seemingly record time, she was ready for service by 29 February.

During the summer of 1930, when the Depression had begun to be felt in Britain, *Majestic* was sent across the Atlantic to operate cruises out of New York. These were three-and-a-half-day mid-week excursions to Halifax, Nova Scotia, or on 'cruises to nowhere', in reality 'booze cruises' to beat Prohibition and often in partnership with *Olympic*, when both ships had to stay in New York for a seven-day stopover.

White Star was feeling the squeeze from the Depression, and this led to staff cutbacks, resulting in reduced maintenance and upkeep on their liners. Sadly, the outcome was that *Majestic* occasionally entered port streaked with rust. In October 1931, *Majestic*'s accommodation was again revised to first, tourist and third. The year 1930 had witnessed a large decline in passenger traffic on the North Atlantic, which by 1931 amounted to a virtual collapse, the receipts for the westbound traffic alone having decreased by 240,000. The unpleasant fact emerged that British lines, including White Star, were losing more passengers than their Continental rivals. Prospective clientele were booking aboard the latest liners in service, like Norddeutscher Lloyd's *Bremen* and *Europa*. In fact, by the end of December 1931, White Star showed a financial loss of £450,777.

In the summer of 1932, the Great Britain Olympic Team, bound for Los Angeles, made the five-day Atlantic crossing aboard *Majestic*, continuing for a further five days across the American continent by Union Pacific Railway until they reached Los Angeles. This gave White Star something of a fillip but nonetheless by the end of 1932 she had incurred a loss of £152,045. However, as 1933 drew to its close, there were definite signs that the worst of the slump was over, although there was no significant improvement in the North Atlantic trade, where the competition from the newer, palatial cabin ships had cut into the profits of the more expensive express liners, all of which were running at a considerable loss. White Star's

The former *Majestic* at Southampton's Western Docks on 16 January 1937 after adaptation into the Royal Navy cadet training ship HMS *Caledonia*, to be based at Rosyth. (Adrian Vicary)

shortfall for 1933 amounted to £353,552. On the brighter side, there was the impending merger with Cunard Line in 1934, and by then the total westbound traffic had bottomed out.

On 19 January 1934, *Majestic* became the first liner to use the newly completed King George V Graving Dock at Southampton. Later, in July 1934, *Majestic* became part of the larger, merged Cunard White Star Line Ltd, which was formed on 1 January that year to take over the trade of the former rival companies. Now, at last they were partners for all future services, although the merger effectively brought an end to the White Star Line. *Majestic*, along with *Olympic*, *Homeric*, *Georgic*, *Britannic*, *Adriatic*, *Albertic*, *Laurentic*, *Doric* and *Calgaric*, were White Star's contribution to the new company. Upon the merger, White Star's Australian and New Zealand interests, represented by *Ceramic* and *Ionic*, were relinquished to Shaw Savill & Albion. *Majestic* remained the world's largest liner and she was the biggest ship in the combined fleet.

Notwithstanding her size, during September 1934, she encountered a huge Atlantic wave that struck the liner forcefully and poured tons of water onto her upperworks. Her master, Commodore Edgar J. Trant, was badly injured and the staff captain had to assume command for the rest of the voyage. Presumably arising from his injuries, Commodore Trant never sailed again. Later in the same month, during the spring tide season, *Majestic* missed the tide and ran aground on the Bramble Bank off Calshot, in The Solent. Fortuitously, she was eventually floated on the next tide under her own power and without damage.

At first *Majestic* formed part of the North Atlantic express service in company with *Mauretania*, *Aquitania*, *Berengaria* and *Olympic*, but on 27 March 1935 *Olympic* made her final round trip from Southampton, after which she was withdrawn from service. From this time, *Majestic*, which by then had replaced *Mauretania*, remained on the run, maintaining the express service along with *Aquitania* and *Berengaria*. The service was operated until the summer season.

When French Line's *Normandie* entered service on 20 May 1935, she became not only the fastest liner in the world, capturing the Blue Riband from Italy's *Rex*, but also at 79,280 gross tons, the world's largest liner. *Majestic* had lost its heavyweight crown, being relegated to second place. *Majestic*'s 207th and final voyage took place from Southampton to New York on 13 February 1936, later departing from New York on 21 February to arrive back in Southampton via Cherbourg for lay-up. The new *Queen Mary* took her place on the New York run when she entered service on 27 May 1936. It is estimated that during her career *Majestic* had steamed some 1,250,000 miles.

With all the hyperbole surrounding *Queen Mary*'s maiden voyage, these events overshadowed *Majestic*'s disposal. After three months' lay-up at Southampton, she was sold for scrap on 15 May 1936 to Thos. W. Ward, for £115,000. Subsequently she was stripped of most of her external fittings, and her masts and three funnels were cut down to facilitate her passage under the Forth Railway

The end of an era. Though the last of the three liners to be completed, *Caledonia*, ex *Majestic*, leaves Southampton for the last time on 8 April 1937: the passing of the first of Ballin's great trio. (Smith Archive / Alamy Stock Photo)

Bridge to the breaker's yard at Inverkeithing. However, around this time the Admiralty Board had decided that a training facility for new entry boy seamen and artificer apprentices should be sited at Rosyth in Scotland. As a purpose-built shore establishment would take some time to erect, a stopgap measure was sought. The huge *Majestic* was ideal for a floating training establishment because she could accommodate the 1,500 new-entry seamen and 500 apprentices, so the Admiralty decided to acquire her. It is believed that the Royal Navy could not buy the liner even at her scrappage value, so an exchange was brokered. The Royal Navy would instead trade twenty-four old warships for scrap whose combined tonnage was equivalent to that of *Majestic*. The exchange was agreed and *Majestic* became the property of the Royal Navy. Under Admiralty supervision, John I. Thornycroft of Southampton then took her in hand to carry out the extensive alterations required.

An intensive programme commenced in which some 2,000 shipyard employees worked for up to thirteen hours a day, seven days a week, for almost eight months to complete the conversion. In this, the cabin-class dining saloon was turned into the boy seamen's mess and the oak-panelled lounge was converted into a gymnasium complete with a boxing ring along with other recreation amenities. *Majestic*'s former elegant Pompeian swimming pool became the baths on board and was one of the luxury fittings retained from her liner days. In addition to these alterations, she was also fitted with seven guns and range-finding equipment. By the time she emerged as the newly refurbished HMS *Caledonia*, the conversion cost to the Royal Navy was an estimated £472,000.

The new training ship left Southampton for Rosyth on 8 April 1937 under her own power with Captain J.W. Binks in command (the following year he would be the master of *Leviathan* on her crossing from New York to the breakers in Rosyth). On Saturday 10 April 1937, HMS *Caledonia* anchored just below Inverkeithing until the tide was low enough to allow her safe passage under the Forth Bridge, aided by eight tugs. Once under the bridge, she was inched into the dockyard basin, where she was to remain for the next two years. All her machinery remained intact and *Caledonia* was capable of producing her own heat, light and domestic hot water. However, fresh water had to be piped aboard and her sewage disposal system was linked to shoreside.

Cadets parade on the quay alongside HMS *Caledonia* on the day of her commissioning, 23 April 1937. In the fifteen days since her departure from Southampton, she has been given a pristine new coat of paint. (HMS *Sultan*)

On St George's Day, 23 April 1937, she was officially commissioned as the fourth HMS *Caledonia*. As such she was the Royal Navy's largest warship in commission and still ranked as the then third largest passenger liner ever built, with a capacity for 2,500 personnel. On 1 September 1939, Britain declared war on Germany and a day later, for safety reasons, the boys and apprentices were moved from *Caledonia* to temporary shoreside billets. Because of her size, the ship was considered to be a potential hazard to the dockside, especially if she was bombed. Hence, during the same month, she was towed out of the basin to a point west of the Forth Railway Bridge and a good distance away from the main shipping channel. She then had water pumped into her bilges so that she settled on an even keel on the bottom of the Forth, a precaution in case she was hit during an air raid, to prevent her breaking adrift and obstructing the main channel into Rosyth. It is believed that the Admiralty were considering disposing of her anyway but any thoughts of seriously considering her for use as a troopship were forestalled when, on 29 September 1939, she caught fire and was gutted.

In March 1940 the Royal Navy sold her back to Thos. W. Ward for scrapping at Inverkeithing, the company beginning the job of salvaging the remains of HMS *Caledonia* during October 1942. They lightened her by removing the superstructure and hull of the ship down to the main deck, which was one deck below the gunwales, then down to the waterline, leaving just the forepeak intact to assist towing. Once lightened sufficiently, the remains of her hull were raised on 17 July 1943 and towed by three tugs to Inverkeithing, where the hulk was beached. During beaching, the last 250ft of her stern section broke away. This was salvaged and towed away to Dalgety Bay, where it was beached and finally broken up during 1943. Apparently, when the bottom plates were cut out from between the webbings, the material, which was of high quality, was found to be in remarkably good condition. So much so the plates were deemed suitable for further use in the construction of new tonnage – much-needed recycled metal at a time of wartime shortages.

* * *

Albert Ballin, the pioneer who had conceived these three great super-liners, died in November 1918, never able to see his ships fulfil his dreams.

TWO

GLORY RESTORED, BUT SHORT-LIVED: NORDDEUTSCHER LLOYD'S *BREMEN* AND *EUROPA*

Through massive losses sustained in the First World War as well as the requisition of many of its big ships by the United States when it entered the war and, subsequently, by the Allied powers under the terms of the Treaty of Versailles, including of course *Imperator* and her near-sisters, Germany's merchant fleet had been all but eradicated by 1920. Largely reduced to coastal shipping alone, it lacked any ocean-going passenger capability with the exception of the old *Deutschland*, by then adapted for cruising as *Victoria Luise*.

This situation eased slightly when, in an agreement with the British to relax the restrictions on German shipping, Norddeutscher Lloyd was permitted to complete and commission a large liner then under suspended construction at the Schichau-Werke GmbH shipyard in Danzig. Started prior to the outbreak of war, by 1920 it was only complete to its double bottom. Originally named *Hindenburg*, this 32,354 gross ton ship entered service in 1924 with the name *Columbus*. A sister ship of that name had already become White Star's *Homeric*.

The emergence of *Columbus* acted as a catalyst for plans to introduce two larger, grander and faster liners, *Bremen* and *Europa*. The 1920s were boom years on the North Atlantic prior to the Great Depression, and Norddeutscher Lloyd was keen to get a piece of the action. The Johnson-Reed Act of 1924 had seriously reduced the immigration quotas for entry into the United States but by way of compensation there was a surge of tourist travel, principally by the American middle classes. In the 1920s, American passengers accounted for 80 per cent of the total traffic. Accommodation grades on liners were modified to cater for this new category of passenger, with tourist and tourist third replacing the old third and fourth classes. The first hint that much larger ships could once more be introduced and operated viably was substantiated by the appearance of *Ile-de-France*, which spurred the construction of new purpose-designed ships.

The new German liners were to be five-day ships across the Atlantic, which, even allowing for the additional passage to and from Bremerhaven through the North Sea, would nonetheless ensure a more frequent regular sailing schedule. The service speed of 27 knots required to achieve this rota demanded total engine power output of around 135,000 shp. In turn, to accommodate such engines necessitated large ships of a minimum of 50,000 gross tons.

It has been stated repeatedly, though without corroborative evidence, that the construction of these two trail-blazing and innovative liners was subsidised by the German government. However, as the late Arnold Kludas, the former leading authority on all German shipping matters, pointed out, that was most definitely not the case. Funding of the liners came in part from Norddeutscher Lloyd's own operating capital, plus credit funds raised through the financial markets, along with compensation payments to be received from the United States in respect of the seizure in 1917 and subsequent retention of numerous German liners. Unlike Great Britain, which had a valid claim under the Treaty of Versailles for

Bremen on the slipway at the Deschimag Weser shipyard, revealing her revolutionary bulbous bow, a structural feature – nowadays usually more extended in length – that featured on the majority of the super-liners subsequently built. (Deschimag AG Weser)

On the slipway at Blohm & Voss, *Europa* prior to her launch. (Blohm & Voss)

recompense for the sinking of *Lusitania* and *Britannic* (which took the form of *Imperator* and *Bismarck*), the United States had not actually lost any of its own passenger vessels and, consequently, did not have the same right to restitution. Therefore, Norddeutscher Lloyd was to be allocated $27.2 million compensation, approximately 100 million Reichsmarks, a substantial contribution to the estimated 68 million Reichsmarks that each of the two new liners would actually cost to build. However, by 1930, Norddeutscher Lloyd had received only $7.35 million (less than 25 million Reichsmarks) and no more compensation would be forthcoming. By 1932 the German government had negotiated the moratorium of the injunction under which it was required to make monetary reparations. As the remuneration paid to Norddeutscher Lloyd had been linked to the country's level of debt, the subsequent abrogation of the reparations obligation led to the termination of all compensation. To make matters worse, the world financial crisis in the wake of the Wall Street Crash in October 1929 added to Norddeutscher Lloyd's funding difficulties, so that the company was obliged to seek additional bridging credits. The German government did no more than provide security for these and the earlier loans.

Launch of *Europa* on 15 August 1928. (Hapag-Lloyd)

Norddeutscher Lloyd entrusted the design of *Bremen* and *Europa* to Blohm & Voss shipbuilders under the leadership of Bruno Paul. While the power plant arrangement for *Bremen* became the responsibility of Dr Ing. Carl Gustav Bauer, that of *Europa* was designed by Dr Ing. Hermann Frahm. In a bid to synchronise construction and have the pair enter service simultaneously, the orders were split. That for *Bremen* was awarded to the Deutsch Schiff & Maschinenbau AG (Deschimag) shipyard at Bremen on the River Weser, as Yard No. 872, while Blohm & Voss obtained the order to build *Europa* at its River Elbe shipyard, Hamburg, as Yard No. 479. The contracts were signed in December 1926.

Keel laying of the two ships thus occurred on consecutive days, on 17 and 18 June 1927. Likewise, the launching ceremonies also took place a day apart; *Europa* on 15 August 1928 and *Bremen* on 16 August 1928. In the event, though, the grand plan for their contemporaneous commissioning was thwarted, falling victim to fate, for on 26 March 1929 a devastating fire engulfed *Europa* during fitting out. The planned joint entry into service, scheduled for July 1929, had to be abandoned and effort turned to salvaging the fire-ravaged *Europa* and, if possible, undertaking her restoration. There had been a danger, with so much water poured into her, that she would list to starboard and capsize at her berth but, fortuitously, she sank on an even keel, settling into the mud of the River Elbe as the tide ebbed. *Europa* was pumped out and raised on 10 April 1929 and, amazingly, despite the extent of the damage, including to her turbines – so extreme in fact that disposal for breaking up had been seriously considered – she was fully repaired within ten months and ready to commence sea trials on 22 February 1930.

The fire had broken out during the early hours in several places, suggesting arson, but the cause was never definitely established and no arrests were made.

Meanwhile, *Bremen*, which it had been intended should be the second to enter service, albeit by only days, was destined to be the lead ship, making her maiden sailing from Bremerhaven to New York via Southampton on 16 July 1929. Whether or not it was for this reason, *Bremen* proved to be the more popular of the pair, although this was commonly the outcome between liner consorts.

Owing to their generally homogeneous design, the ships' technical specifications were similar. Gross tonnage of *Bremen* was 51,731, while that of *Europa*, as built, was 49,746, later increased to take her over the 50,000-gross-ton super-liner demarcation limit. Contemporary postcards gave her tonnage as 51,000 gross. *Bremen*'s main dimensions were 938ft length overall and 102ft across her beam. *Europa* was longer at 941ft with an identical beam measurement. They had eleven decks: sports & sun, upper promenade, main promenade and A to H. Of these, decks B to E extended from stem to stern. Four sets of high-speed steam turbines, rated at 105,000 shp, were coupled to quadruple screws via single-reduction gears, ensuring a service speed of 26.5 knots in both cases.

Passenger accommodation on *Europa* was 860 first class, 500 second class, 305 tourist class and 617 third class. For *Bremen*, the breakdown was 811 first, 500 second, 300 tourist and 617 third class: only minor differences between them. The crew complement of both ships was 965. Reclassification of the accommodation in 1936 resulted in a three-class arrangement of cabin, tourist and third class.

For Germany, given the parlous state of the nation's economic circumstances and the fact that just ten years earlier its merchant marine had been virtually non-existent, the conception, realisation and introduction of *Bremen* and *Europa* on the North Atlantic run was an extraordinary achievement. At the time there was nothing to rival them, not only in their performance but in the ground-breaking nature of their design both internally and externally. The French liner *Ile-de-France*, introduced to the Atlantic passenger service in 1927, had wowed the maritime world with her haute modern art deco interior decoration and appointments, but in her exterior styling she remained irrefutably a link with the past. Her three tall, slim funnels and her counter stern, square bridge front and barely raked stem were all hangovers from the Edwardian era. Credit for introducing a modern phase of ocean liner design, a radical departure from the conventions of the day, indubitably

Europa completely ablaze at the fitting-out quay on 26 March 1929, swept by the fire that came close to preventing her from ever entering service. (Hapag-Lloyd)

The fire damage was extensive. Her upper structure collapsed, as seen here, and much of her interior was gutted. (Hapag-Lloyd)

Bremen arrived at Southampton from Bremerhaven on 1 July 1929 and proceeded directly into the 60,000-ton floating dock to have her hull cleaned and painted. She was the first liner flying a foreign flag to be accommodated in this dock. The men standing on the dock floor give a good impression of the ship's immense size. (Historical Southampton)

belongs to *Bremen* and *Europa*. They exhibited features that would not have been out of place more than half a century later.

Their public rooms were light, bright and stylish, perhaps more Bauhaus than art deco, a style that was actually described as Moderne Sachlichkeit (modern pragmatic). The suites and cabins also offered numerous modern amenities but the *coup de grâce* was their streamlined, sleek exterior lines. They had broad, squat funnels, rounded corners to the superstructure, recessed anchors and a revolutionary hull form with a hollowed bow and a raked stem extending down to a bulbous forefoot, all designed to improve them aerodynamically and hydrodynamically. It wasn't just streamlining but exceptional engineering, the first time that performances were to be enhanced by hull design rather than just by uprating engine power output.

Bremen at Bremerhaven. She may be en route from her builder's shipyard to the Columbus Quay, or possibly it is the occasion of her maiden voyage on 16 July 1929. (David L. Williams collection)

A novel feature on *Bremen* alone was the overlapping of part of her underwater hull plates so that the edges faced forward rather than aft, a bizarre configuration that seemed to contradict logic. Not so. It was another carefully thought-out arrangement, confirmed by model tests prior to the commencement of construction. In attaching the plates in this fashion, water running along the hull sides encountered a series of blunt butts. Ordinarily, this would increase skin friction and resistance but at high speeds it did the opposite, permitting a fraction of a knot in speed to be gained. Similarly, the streamlined rudder also worked to the same end.

As had been done with two of the *Imperator*-class ships, the exhaust trunking from the boilers was divided to run up the sides of the hull, reconnecting just beneath the funnels.

Yet another distinctive feature of *Bremen* and *Europa* was an aircraft cradle and catapult mounted on the sports deck, just abaft the forward funnel, from which a Heinkel He 12 floatplane was launched to make deliveries of mail hours before the ships themselves docked. They were also the first liners to introduce the practice of having their illuminated names displayed amidships on either side of the upper promenade deck.

Operational experience, especially from *Europa*'s maiden voyage, revealed that the funnels of *Bremen* and *Europa* were, in fact, too squat. Passengers on the aft decks complained about having oily, sooty smoke smuts descend upon them. Executing a quick remedy, *Europa*'s funnels were raised in height by 15ft and those of *Bremen* were adjusted likewise soon after. As a matter of interest, this revealed one of the distinguishing features between the pair. The locations of the grilled openings of the engine room ventilation shafts were located at the front of the funnels on *Bremen*, whereas on *Europa* they faced aft. The plan form of their funnels was also dissimilar, those of *Bremen* being pear-shaped while on *Europa* they were elliptical.

Under the command of Captain Leopold Ziegenbein, *Bremen* secured the Atlantic Blue Riband on her maiden Atlantic crossings, with average speeds of 27.83 knots westbound and 27.92 knots eastbound. *Europa* also took the westbound record on her maiden voyage, which commenced on 19 March 1930, achieving an average speed of 27.91 knots, completing the crossing in four days and seventeen hours. She never held the Atlantic Blue Riband for the eastbound crossing but *Bremen*, proving herself to be the faster of the two, improved her performance in July 1933, wresting the record from *Europa* with a speed of 28.51 knots. *Bremen* retained the eastbound record until it was lost to *Rex* in August 1933 and the westbound record until it was lost to *Normandie* in June 1935.

Apart from their record-breaking crossings, *Bremen* and *Europa* settled into a steady and reliable, if unspectacular, routine, calling at Cherbourg and Southampton en route to New York as well as on the return leg to Bremerhaven. Averaging eighteen or nineteen round voyages each season, they ranked with the best on the Atlantic run, although their passenger numbers declined in common with the experience of all other transatlantic liners in the austerity of the early 1930s. Passenger returns were only beginning to register improvement in the late 1930s when, as if 'rain had stopped play', the outbreak of the Second World War intervened. The only liner to better them over the period from 1936 to 1939, in both round trips and passenger numbers, was Cunard's *Queen Mary*.

Opposition to the Nazi regime led to a number of protest incidents in New York, during one of which, in July 1935, the Swastika ensign at the stern of *Bremen* was torn down. However, newspaper accounts suggest that these demonstrations had little impact on the passengers who chose regardless to travel aboard the Norddeutscher Lloyd pair.

Bremen made two world cruises, the second in early 1939, transiting the Panama Canal with little room to spare, just 4ft on either side in the canal lock chambers and, in so doing, effectively setting what has since come to be known as the Panamax standard, the broadest beam that could pass safely through the locks.

The onset of war found *Europa* alongside in Bremerhaven after completing her last pre-war voyage from New York. It is suggested that, in the haste to return to her home port, *Europa* did not make her customary call at Southampton but steamed past the Isle of Wight, taking Britain-bound passengers with her. As for *Bremen*,

Port bow close-up view of *Europa* at the shipyard when completed and ready for delivery to her owners. (Blohm & Voss)

she was making her final westbound crossing. Her return leg from New York was to be heralded as an epic and daring voyage, taking her far from her normal route to evade the unwelcome attention of patrolling British warships.

When *Bremen* was some 900 nautical miles out from New York on the westbound passage, her Master, Captain Adolf Ahrens, had received a radio message ordering him to turn about and return to Germany because the outbreak of war was imminent. The order was ignored and *Bremen* continued on her way, to arrive in New York on 28 August 1939. Once there, Captain Ahrens wanted to make a quick turnround to depart with the minimum of delay. His concern was to avoid having his ship interned in America as had happened to *Vaterland* twenty-two years earlier, but clearance to sail was refused and zealous US Customs officials held her back while they conducted a protracted and thorough search. According to the rules of neutrality, America was required to check for any contraband stowed aboard – weapons and ammunition that could permit a vessel to convert at sea into a commerce raider, or any of a number of illicit materials that could be used in the production of munitions. But, for *Bremen*, the inspection was dragged out inordinately. For balance, though, it must be said that *Normandie* and *Aquitania* were also subjected to comparable treatment.

The prolonged examination at least gave ample opportunity for the ship to be fully refuelled and finally, on the evening of 31 August, *Bremen* was permitted to sail. Captain Ahrens' fear was that *Bremen* would immediately run into blockading Royal Navy warships, alerted to her flight by the US authorities. But he had been devious, earlier transmitting a message back to Germany implying he would make a dash for Havana and, if intercepted en route, he would scuttle his ship rather than allow it to be captured.

In fact, as soon as she entered international waters, *Bremen* was steered in a northerly direction, heading for the Arctic Circle. Her crew was charged with making preparations in case scuttling became necessary, as well as with the painting of the ship in shades of grey to help conceal her in the hazy conditions of the North Atlantic. All running lights were extinguished, deadlights were fixed over all portholes and numerous lookouts were posted around the ship. There were indeed units of the Royal Navy hunting for *Bremen* but no one seriously believed she would be taken so far north, through the Denmark Strait, south of Greenland, into a sea area where she ran the risk of collision with an iceberg.

For *Bremen*'s crew, lacking the kind of clothing required for such northerly climes, it was a miserable and uncomfortable ordeal, but they shared their Master's determination to get their ship home. Their concerns were heightened when in mid-ocean, on 3 September, the ship was notified of the commencement of hostilities with Great Britain.

As completed, finally entering service on 19 March 1930, *Europa* had unusually short funnels. While they added to her sleek appearance, they also caused a smoke nuisance to passengers on her aft decks, as hinted at here. Designers thought that smuts from oil-fired boilers would not fall on the after decks. They were wrong! (Hapag-Lloyd)

Only eleven days earlier, Germany had entered into a non-aggression pact with the Soviet Union and the assistance of the Russians was now called upon to permit *Bremen* to reach the initial safety of Murmansk. With approval granted, *Bremen* proceeded north-east to pass south of Svalbard and into the Barents Sea, finally arriving at Polarnöye in the Murmansk Oblast on 6 September, having sailed 4,045 nautical miles in six and a half days, maintaining a speed of 27.5 knots. Twelve days later, 870 of *Bremen*'s crew returned to Germany via a lengthy train journey, leaving only around eighty officers and crew to keep the ship in good order and prepare for the final part of her odyssey. Later, when it came to the time to undertake the ultimate part of her mission, the dash for home, the return of thirty-five members of *Bremen*'s crew from Germany was requested, mainly seamen

Another feature of the Norddeutscher Lloyd super-liner pair was an aircraft catapult for launching the mail thirty-six hours before the ship arrived in port. Here the Heinkel He 12 floatplane is warped aboard *Bremen* on 4 October 1931. (Hilary de Kerbrech)

but also engineers and navigating officers. In the event, the crew numbers were supplemented by a total of fifty-seven men, included among them a number of Kriegsmarine personnel. During her three-month sojourn at Murmansk, fuel had been replenished by the German tankers *Wilhelm A. Riedemann* and *Friedrich Breme*, which had evaded the British blockade.

The intention was for *Bremen* to make a break for the open sea, heading southwards and hugging the coast of Norway, before entering the North Sea and skirting Denmark to eventually reach the safety of her home port. Some consideration was given to entering the Baltic to make for a north German port but this option was abandoned, with the preference being to get the ship to Bremerhaven. At the time, both Norway and Denmark were neutral, which was considered advantageous to the homeward dash, and the closer *Bremen* approached Germany, the more likely it was that she would be able to benefit from a measure of protective air cover.

Captain Ahrens chose his moment to make a run for it, perilous though it would be, exploiting the darkness and low visibility of the Arctic winter. Early on the morning of 7 December, protected by a snow squall, *Bremen* put to sea. Battling stormy seas in low light, all went well until, on 12 December, she was intercepted by the British submarine HMS *Salmon* engaged in patrols along the British blockade line. The submarine instructed *Bremen* to surrender. The tense exchange that ensued was abruptly ended by the timely intervention of a Dornier flying boat despatched from Germany, which compelled *Salmon* to crash dive. After diving, the submarine's commanding officer, Lieutenant Commander E.O. Bickford, declined to sink the liner by torpedo, believing her not to be a legal target. Thus, *Bremen* was permitted to make her escape.

Greeted by a tumultuous welcome, *Bremen* reached home and was secured alongside the Columbus Quay, Bremerhaven, on 13 December 1939. She had written one of the great stories of maritime derring-do. Yet, unknown on that day of exuberant congratulations, it was to be her last voyage for, apart from some repositioning, *Bremen* was fated to never make another ocean crossing.

Bremen seen underway before the height of her funnels had been increased. (Arnold Kludas)

At speed, *Europa* is seen after her funnels had been heightened. (Norddeutscher Lloyd)

Leaving the New or Western Docks, *Bremen* sails from Southampton. (David Hutchings collection)

Laid up together in Bremerhaven, *Bremen* and *Europa* served as naval accommodation ships. Later, in the summer of 1940, they were activated to participate in the planned invasion of Britain, Operation Seelöwe (Sealion). At least, that was the supposed intention, for the reality was somewhat different. Moved to Hamburg, they were taken in hand for adaptation into military transports, numbers 801 and 802. Both ships were painted in dazzle-type camouflage colours and it was said that large apertures were cut in the sides of their hulls to permit rapid embarkation and disembarkation of troops. But it was all part of a sophisticated ruse, a simulated diversionary tactic, cooked up by General Admiral Rolf Carls of the German Naval War Staff and intended to draw British attention away from the real invasion force.

Code-named Operation Herbstreise (Autumn Journey), the plan was to give the impression that the actual invasion was taking place along the British coast between Edinburgh and Newcastle, with the fake landings targeted for Blyth. The purpose was to lure away and tie up part of the Royal Navy's Home Fleet engaged in patrolling the British blockade between the Faroe Islands and Iceland. Four 'dummy' convoys would set out simultaneously from the area between Bergen and Kristiansand in the north and the German Bight in the south, three from Norwegian ports, now in German hands, the last combining units leaving from Cuxhaven and Bremerhaven.

Convoy IV, of which *Bremen* and *Europa* would form a part, was also to comprise the former ocean liners *Gneisenau* and *Potsdam*, the light cruiser *Emden* and five torpedo boats, plus, for good measure, a small hospital ship. On invasion day, the convoy would depart at 0830 hours and continue in a north westerly direction until 2130 hours, by which time it would be halfway to Scotland, when it would turn about and return to base. If necessary, the manoeuvre would be repeated the following day. Meanwhile, the genuine Sealion invasion force would be crossing the Dover Strait to assault Britain on the shores of Kent.

As part of the subterfuge, intensive troop movements would take place in Denmark and southern Norway, all hopefully spotted by RAF reconnaissance aircraft. The loading of troops aboard *Bremen* and *Europa* would be simulated at Wesermünde although, in fact, they would put to sea completely empty, whereas troops would board *Gneisenau* and *Potsdam* at Hamburg only to be offloaded at Cuxhaven before the ships entered the North Sea.

It was a convoluted plot, said by some German military officials to have grown to such proportions that it overshadowed the invasion operation proper. In the event, the Luftwaffe failed to dominate the skies over the English Channel and southern Britain, a fundamental precursor to launching the landings, and so both Operations Seelöwe and Herbstreise were abandoned. Set for launch on 16 September 1940, for all practical purposes it was cancelled that very month although officially it continued to be a projected campaign until February 1942.

There ensued a period of lay-up in Bremerhaven as the Kriegsmarine had no requirement for large ocean-going troopships, virtually all of Germany's conquests having been made across land borders. This idle sojourn came to an abrupt end for *Bremen* on 16 March 1941 when she was swept by fire. The inferno raged for two days while the local fire service swamped her with water. As with *Europa* twelve years earlier, arson was immediately suspected due to the speed with which the

A view of *Bremen* from her starboard bow after her funnels had also been raised in height. (Authors' collection)

Showing off her clean modern lines, *Bremen*, here seen steaming at speed, introduced a step advance in modern ocean liner design along with her sister *Europa*. (Hapag-Lloyd)

ship had been overwhelmed. Top heavy with water, *Bremen* adopted a list, leaning over to the quayside on her starboard beam. Once extinguished, inspection revealed *Bremen* to be realistically beyond recovery and, as it was wartime, little was done besides getting her on an even keel. Bremerhaven was not a naval port so *Bremen*'s charred hulk was not obstructing a valuable strategic resource.

The true cause of the fire has been a matter of opinion and conjecture ever since, various elucidations having been proffered. Some thought the fire had been started by strikes from bombs dropped by the RAF in a raid on Bremerhaven a short time earlier. Certainly, more than once, Allied warplanes had attempted to destroy *Bremen*, and the nearby *Europa* had been slightly damaged the previous November.

Bremen approaches one of the Panama Canal locks during her 1939 cruise. (Sueddeutsche Zeitung Photo / Alamy Stock Photo)

It has also been claimed that a truculent junior deck-hand was responsible, having started the fire in an act of resentment, aggrieved at a ticking-off by an officer or senior crew member. This lad, 15-year-old Kurt Walter Schmidt, had apparently confessed to the crime and, subject to the summary justice then being meted out, he was executed.

However, author Peter Huchthausen, US Navy Retired, a former naval attaché in several eastern European countries, presented an alternative explanation following lengthy research at the Deutsches Schifffahrtsmuseum, Bremerhaven. The suspected culprit he identified was a Gustav Schmidt, a former member of *Bremen*'s crew, who was arrested in Constanta, Romania, after bragging that it had been he who had set *Bremen* alight in an act of sabotage. Witnesses claimed that he had been seen in covert discussions with two foreign men who, it was suggested, had paid him more than 300,000 Reichsmarks to carry out the deed. Though a trial never proved his guilt, the story goes that he too received a death penalty verdict. But were Kurt Walter Schmidt and Gustav Schmidt one and the same person?

So many years after the event, the true identity of the perpetrator or perpetrators behind the tragic loss of *Bremen* may never be revealed irrefutably. But there remains one last possibility, the papers that recorded the post-surrender interviews with SS Major Horst Kopkow, responsible for the Gestapo Security Police in the Bremen area. They reveal, it is claimed, that he carried out the investigation into the fire on *Bremen* and in them he may well have identified the culprit's true name.

As for her remains, she was systematically stripped down to the waterline where she lay, removing quantities of metal to be recycled in Germany's munitions factories, which by then were beset by material shortages. Final wreck clearance did not get under way until after the war's end. Even if she could have been recovered – a dim prospect in the wartime circumstances – there would have been no use for her. In 1946, the occupation authorities, the US Army Europe, which on 7 June 1945 had established the 17th Major Port

(Bremerhaven Port of Embarkation), ordered the removal of the partially demolished hulk and it was towed across the river and beached on a sand bar at Blexer Reede, Blexen, on the Nordenham bank of the Weser. Using explosive charges, the dismantling and demolition of her discarded lower hull proceeded there through to 1956. The residual remains, parts of *Bremen*'s double bottom, can still be seen to this day at low tide.

Meanwhile, *Europa* languished, likewise having no obvious military value, other than as quarters for service personnel, until a year later when the aerial threat at sea from British aircraft carriers prompted the Kriegsmarine to pursue the commissioning of carriers of its own, leading to her becoming a focus of interest. One would have thought the first priority would have been to expedite completion of the part-built, purpose-designed carrier *Graf Zeppelin*, but instead several ocean liners were earmarked for conversion, among them *Europa*, consuming valuable resources on these projects.

Planning for *Europa*'s conversion commenced in May 1942 under the designation Flugzeugträger I (Aircraft Carrier I). Her length would have been increased to 956ft to provide an extended flight deck, her maximum beam raised to 121ft by the addition of bulges for improved stability and she would have had a full load displacement of 56,500 tons. Multiple anti-aircraft batteries would have been installed in thirty-two double mounts and up to thirty-six quadruple mounts, and her aircraft complement would have been eighteen Junkers Ju 87c dive bombers and twenty-four Messerschmitt Bf 109t fighters, both variants suitably adapted for naval service.

Aircraft Carrier I would have been one of the largest at that time. However, the conversion did not proceed because a number of critical structural issues were encountered, primarily arising from the need to incorporate a 709ft by 82ft to 98ft hangar that would have weakened her primary structure, already considered vulnerable. Stability was another issue. Even with an absence of armour – obviated for that very reason – it was thought that it could

In this view, *Bremen* can be seen passing through the Gatun Lock on 15 February 1939. (US National Archives 185-G-0817)

Bremen seen arriving at Bremerhaven from Polarnöye, near Murmansk, on 13 December 1939, following her secretive dash south along the Norwegian coast. (L.L. von Münching)

Bremen and *Europa* painted in camouflage colours moored alongside the Columbus Quay, Bremerhaven, in 1940. *Europa* is the further ship of the two. (David Hutchings collection)

Listing to starboard against the quay, *Bremen* is seen engulfed by the fire that destroyed her in this poor-quality newspaper image. (AGW Archive)

be potentially compromised anyway by the proposed modifications. Besides those disquietudes, her fuel consumption was also regarded as excessively high, limiting her duration and range at full speed. Fundamentally, though, in a typical Nazi turf war, the Luftwaffe was against the adaptation of existing aircraft types and the provision of sufficient, experienced pilots.

So *Europa* saw out the war much as she had spent it for the previous two years, laid up inactive at Bremerhaven. Tied up alongside the Kaiserhafen III quay, she was seized by American troops as a war prize when they entered the port on 8 May 1945. Apparently, the intention had been to scuttle her but the order was rescinded with the collapse of the Nazi regime. Instead, after a hasty overhaul in the Bremerhaven drydock, she was pressed into service as the troop transport USS *Europa*, AP-177, commissioned on 25 August 1945. Although she completed just three trooping voyages, she received citations and Occupation Service Medals for two periods of service. The first voyage commenced on 11 September 1945 when she cleared Bremerhaven to pick up 4,500 US troops at Southampton for return to New York. On her arrival, on 26 September, she was drydocked at Bayonne, New Jersey, where she was taken in hand for adaptation to permit her to carry up to 6,600 troops. All but four of her lifeboats were removed and replaced by Carley Floats with a capacity for 9,000.

The work complete, she sailed for Southampton on 11 November 1945 and with another full complement of repatriating troops arrived back in New York fourteen days later. Her next round voyage to Southampton commenced on 1 December, returning on 18 December. A number of issues, electrical and plumbing faults, small fires and even cracks in her structure, resulted in the transfer of USS *Europa* to the Todd Shipyard in Hoboken for further attention. She was back at sea on 15 March 1946 bound for Bremerhaven with a call en route at Kirkwall in the Orkney Islands. However, despite the considerable amount of time and money invested in her, the US naval authorities had no desire to retain her, and on her arrival in her old home port on 2 May 1946 she was decommissioned.

Following the end of the Second World War, the occupying forces moved *Europa*, still wearing camouflage, on 28 June 1945 from the Kaiserhafen III, where she had been laid up, to the Bremerhaven drydock in readiness for adaptation for troop repatriation duties. (US National Archives 306-NT-3043V)

The troop transport USS *Europa* (AP-177) sails from the River Weser on 11 September 1945 at the start of her first trooping voyage. (US National Archives 80-G-0352259)

Renamed *Liberté*, the former *Europa* moored alongside a mole in the Bassin Thèophile Ducrocq, Le Havre, after her transfer there from Cherbourg. (French Line)

She was transferred to the War Shipping Administration, and the United Nations Reparations Commission ultimately awarded her to France in accordance with the Blum-Byrnes Agreement, signed on 28 May 1946 by Léon Blum and James F. Byrnes, as a partial compensation for the loss of *Normandie* in New York four years earlier. In fact, the French government had been lobbying for *Europa* since late 1945, before the US Navy had rejected her, but it was thought then that their petition had failed and she was intended, instead, for Great Britain. A letter from Henri Bonnet, the French Ambassador in Washington, to the Minister of Foreign Affairs in Paris, Georges Bidault, sent on 11 January 1946, implied that with interest in her retention by the US waning, the Reparations Commission was intending to offer *Europa* to England, presumably to offset the loss of the Canadian Pacific's *Empress of Britain*, the largest British merchant ship lost in the war.

However, it was to France that she was destined. The offer to the French under the agreement had been a quantity of Liberty ships or *Europa*, or both. Essentially, America was within its rights to disregard the Commission and to trade the ship, even though they had only held her for a year, since it had been their forces that had seized *Europa* as a war prize.

Assigned to the Compagnie Générale Transatlantique (CGT/French Line), *Europa* sailed for Cherbourg on 11 June 1946, already with Transat funnel colours, arriving there three days later. Her destination of Cherbourg instead of Le Havre had been decided upon because Cherbourg was the first French port on the Channel to be liberated and cleared of mines and other German sabotage ordnance, and had a reasonably intact terminal facility. There, after some prevarication as to how she should be rechristened, she was given the name *Liberté*, appropriately enough, in a ceremony on 27 July 1946 attended by government and local and regional dignitaries, officials of the French Line and representatives of the Marine Nationale. The following day she left for Le Havre, her future home port, arriving there on the evening of 29 July, dressed overall and flying the Tricolor, with her new name proudly displayed on her sides, forward and aft. *Liberté*'s upperworks had been painted white but her hull remained in US Navy grey pending the completion of her intended refurbishment.

Destruction of the Gare Maritime Transatlantique and much of the Quai Joannès Couvert during the Allied bombardment of Le Havre prevented ships of all sizes, not only the large ocean liners, from docking there. Consequently, *Liberté* was moored to a jetty connected by a narrow pier to a central mole in the Bassin Thèophile Ducrocq. It was an exposed position, lacking protection from south-westerly winds.

In the immediate post-war period, there were numerous pressing demands in the effort to restore 'normality', as much a burden for France as it was for the many other European countries whose infrastructure had been devastated in the war. But there was an acute shortage of funds and capacity, and priorities had to be set by governments and commercial organisations. Shipping companies had wasted no time in placing contracts for new tonnage, and shipyard order books were crammed. The former *Europa* had to wait her turn and thus she remained tied up at the jetty in the Bassin Thèophile Ducrocq. On 8 December that year her vulnerable situation proved to be the undoing of the luckless liner.

Ravaged by a fierce Atlantic storm, a tempest the like of which locals claimed had not been experienced in decades, *Liberté*'s mooring lines parted and she was swept across the Bassin by the storm-force winds. She drifted some 600ft towards the wreck of the liner *Paris*, which was still lying half-submerged at the extreme southerly end of the Quai Joannes Couvert where it had lain since it was destroyed by fire and had capsized on 18 April 1939. The *Liberté* struck the wreck aft on her starboard side, sustaining a 50ft gash in her underwater hull. It appears that she was then pushed away from the *Paris* by the wind, all the time taking on water and rapidly developing a severe list to port. By the following day it had increased to 45 degrees. Gradually, though, as she flooded up to C deck, she righted herself as she settled on the harbour bottom on an even keel at an acute angle to the quay. It was the liner's second

After tearing her hull open on the wreck of the liner *Paris*, the flooded *Liberté* initially adopted a precarious list. Fortunately … (Jean-Yves Brouard collection)

… she righted herself and settled on an even keel near to the Quai Joannés Couvert, thus facilitating the subsequent salvage and recovery. (David Hutchings collection)

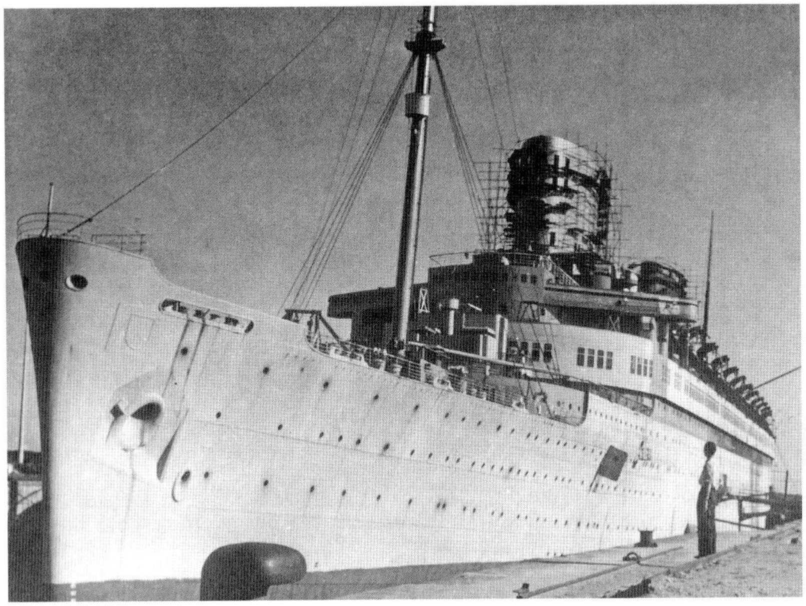

Liberté under reconstruction in the Joubert Lock at the Penhöet shipyard in Saint-Nazaire. (Jean-Yves Brouard collection)

The fully restored and renovated *Liberté* steaming at speed. In many respects she looked little different from how she had as *Europa*, apart from a change to her funnel colours, which had also been given fin-style smoke deflectors. (Jean-Yves Brouard collection)

immersion. Fortunately, there was clearance all around her, which proved to be beneficial to the subsequent salvage effort. The wait to be restored to her former glory had been one issue but now there was the additional task of refloating her, with all its associated costs.

Salvaging the *Liberté* got under way on 15 January 1947. Whatever her future, the waters of the Bassin had to be cleared, along with the reconstruction of the quay and the passenger terminal, to permit shipping services to resume at what was, after all, France's premier passenger port.

Engineers of the French Line's technical bureau, led by Franco Spinelli, studied the structural characteristics of the liner to determine the best method of refloating her. Under their guidance, the task was undertaken by the Compagnie International de Récuperation de Métaux aided by the Les Abeilles towage firm under the command of André Picard.

First, the hull had to be sealed and, with the aid of divers, the breach in her hull plating made watertight. Pipes were installed throughout the ship in preparation for pumping her out. Simultaneously, an anchorage plan was devised involving nine mooring lines each with two 7-ton anchors. A pontoon crane with a lifting capacity of 200 tons was brought to the scene. Meanwhile, a special docking device was constructed for erection alongside the Quai Joannès Couvert, to which it was intended to secure *Liberté* once she was afloat.

Maintenance of the vessel's stability was paramount. To that end, top weight was reduced wherever possible while weight was added to *Liberté*'s lower levels and, to counteract the effect of any residual free water movement during the pumping-out process, partitions were installed to confine it to the centre line of the hull.

Pumping commenced on 29 March 1947, delayed for a day by bad weather. On 1 April, winds of 90mph interrupted proceedings but on that day there was the first hint that *Liberté*'s stern was beginning to float free. The salvage team had to proceed with great caution, however, to ensure the ship remained level and to avoid subjecting the hull to undue stresses. A severe storm on 16 April again punctuated proceedings but eight days later the *Liberté* was finally

Left: Another view of *Liberté* following her entry into service with the French Line in August 1950. She is moored off the Motherbank at Spithead from where her passengers were embarked and disembarked via tenders from Southampton Docks. In June 1953, during her visit to The Solent, *Liberté* represented France at the Spithead Coronation Naval Review for Queen Elizabeth II. (Phil Fricker)

Below: Another view of the regenerated *Liberté*, showing her initial aspect as a French Line ship. (French Line)

and fully afloat. Supported by two pontoons, she was thoroughly inspected to permit a decision on her future.

But for the CGT there was no question, they were fully committed to a complete renovation of the salvaged but seemingly ruined ship. The next task, therefore, was to prepare her for transfer to the Penhoët shipyard at Saint-Nazaire.

By 9 November 1947, she was ready for departure. With propulsion from her own two central propellers – her engines having been partially reactivated – and assisted by six Abeille tugs, *Liberté* left Le Havre at a steady speed of 14 knots. On the afternoon of the 11th, she was assisted into the shipyard by the CGT tug *Titan* with help from *Abeille 4* and *Abeille 25*. A long period of transformation lay ahead, lasting almost three years.

But *Liberté* was not simply restored and reconditioned. Planned for the liner was a virtual rebuild as she was to be born again very much in the French style. Attention was given to every detail as the German '*fräulein*' became unmistakeably a French '*mademoiselle*'. Her appointments, her decorations, every feature assumed a Gallic flair and no expense was spared in this amazing rebirth, the cost of which amounted to £7 million, more than £300 million by today's values.

Regrettably, the luckless liner was struck again by adversity during her reconstruction. A fire broke out aboard her on 18 October 1949 when an acetylene cylinder exploded, severely damaging around sixty first-class staterooms. Serving only to delay completion and postpone her first voyage under her new French owners by less than a month, this additional impairment was swiftly rectified.

How do you describe in just a few lines the extraordinary and superb remodelling of *Liberté*'s interiors? It was a mélange of superlatives in every respect, accounting for the huge expenditure involved. For the French Line, she was the nearest they could reach at that time to having an acceptable replacement for *Normandie*. And not just in a symbolic sense for there were physical reminders of *Normandie* aboard her in the form of decorative souvenirs, pieces of artwork and other displayed artefacts rescued from the great liner when she caught fire in New York in February 1942 (see Chapter 4).

From a safety point of view, every requirement of the 1948 revision of the Safety of Life at Sea (SOLAS) regulations was addressed in her renaissance. Her strength deck was reinforced by doubling the plating for much of its length, the watertight bulkheads raised by a deck and the fore and aft water tanks removed altogether, while the four located amidships were filled with additional ballast. To compensate for the lost fresh water capacity, evaporators capable of producing 600 tons daily were installed.

Liberté's speed credentials were to be a dimension of the past. Four of her original boilers had been removed and her turbines down-rated, reducing engine horsepower. Thus, a revised service speed of 24.5 knots resulted in six-day crossing times for the remainder of her career.

Externally, she did not immediately look so different from how she had as *Europa* apart from her Transat colours and a distinctive feature incorporated on her funnels – aileron-type fins had been fitted as an additional smoke deflection measure, although that was not to be the end of that particular matter. During her annual refit in February 1954, the fins on her funnels were removed and new dome tops of Lascroux design were instated with vents on their leading edge. This increased their height yet again, and it could be claimed that *Liberté*'s funnels were the largest of any ship then afloat.

The fully renovated *Liberté* had a gross tonnage of 51,839 and her revised passenger capacity was 569 first class, 562 cabin class and 382 tourist class. She returned to Le Havre, commencing her maiden sailing as a French Line ship on 17 August 1950, her chequered past largely behind her. She was an instant success and proved to be one of the most popular liners ever to operate the Atlantic run. In a sense she was the epitome of French culture afloat – the style, the décor, the service and, of course, the cuisine. She went on to make 398 Atlantic crossings (199 round voyages) with virtually full passenger lists until her final days. It is understood that on her Southampton calls she never proceeded to the docks but, like *Normandie*, tendered off the Motherbank, Spithead, or in Cowes Roads. Plymouth was initially adopted for her UK calls on the return crossings from New York.

The final profile of *Liberté* after her funnels were modified in 1954 with domed Lascroux-type tops having smoke deflection vents. (French Line)

Liberté was fêted too by the film industry, the shooting of two movies taking place aboard her, making for good publicity. First was *The French Line* in 1953, an RKO film featuring Jane Russell and Gilbert Roland; a year later Paramount used *Liberté* as the set for *Sabrina* starring Humphrey Bogart and Audrey Hepburn.

During her career, *Liberté* completed only two cruises. The first, a twenty-eight-day excursion from New York that commenced on 11 February 1952, took her to the Caribbean and South America with calls at Fort-de-France (Martinique), Kingston, Barbados, Bahia (Brazil), Rio de Janeiro and Nassau, Bahamas on the return leg. Eight years later, in the twilight of her career, she undertook a Christmas cruise, sailing from New York on 21 December 1960. The itinerary included St Thomas, Martinique, Curaçao, Port-au-Prince (Haiti) and Nassau, returning in the New Year of what was to be her last full year in service.

Alongside at the reconstituted Quai Joannés Couvert, Le Havre, *Liberté* in the late 1950s. (French Line)

Ill fortune continued to plague *Liberté* in the guise of tempestuous weather, and more than once she battled through wild seas, arriving in port late. On one occasion, on 29 December 1956, while westbound from Le Havre, she was struck by enormous waves. A hatch cover was smashed and three of the derricks on her foremast were ripped off, such was the force and volume of the water that hit her. The damage she sustained was considered serious enough to warrant return to her home port for repairs before she was able to resume the voyage.

Her final transatlantic voyage from Le Havre commenced on 2 November 1961, departing New York on the eastbound leg eight days later amid a send-off by the Big Apple that reflected her esteem and popularity. She was laid up on her return, her flagship status about to be transferred to the new *France*, which arrived in the port for the first time on 23 November 1961. *Liberté* was to be disposed of, but the French Line was anxious to avoid the kind of debacle that had befallen her former fleet mate *Ile-de-France* when she had been subjected to explosions and a simulated foundering – she was indeed semi-submerged in shallow water – for the Hollywood film *The Last Voyage*. A tentative sale for temporary employment as a hotel ship at Seattle during the World's Fair of 1962 fell through for financial reasons as it would have necessitated towing her through the Panama Canal. Instead, *Liberté* ended her days in the scrapyard, broken up at La Spezia by Terrestre Marittima. Sold on 30 December 1961, she departed Le Havre for the last time on 25 January 1962. Within a year little remained of her.

THREE

ELEVATING THE MEDITERRANEAN ROUTE TO NEW YORK: ITALIA LINE'S STYLISH SUPER-LINER *REX*

Prior to the First World War, the transatlantic passenger shipping services from Italy were relatively minor in scale when compared with those operated out of north-western Europe and the British Isles. Emigrants constituted the main traffic. The largest Italian passenger vessels ordered up to that time were *Duilio* and *Giulio Cesare* of Navigazione Generale Italiana (NGI), the construction of both of which was suspended during the war and did not resume until 1920.

By 1919, through mergers and acquisitions, there were only three major players left operating passenger schedules from either Genoa or Trieste to New York. They were NGI, Lloyd Sabaudo and the Cosulich Società Triestina di Navigazione (Cosulich Line), formerly the Austrian concern Unione Austriaca.

Italy's transatlantic services suffered from a profound disadvantage compared with their northern rivals – distance! It was 4,050 miles from Genoa to New York, whereas it was only around 3,200 miles from Southampton or Cherbourg, slightly longer for the German liners operating out of Bremerhaven and Cuxhaven. In duration of passage, this translated into from five to six days on the northern route against as much as two weeks for the run from Italy by the slowest Italian-flag vessels. However, the realisation that Italy could offer a unique advantage over the other main North Atlantic shipping lines was beginning to register – a crossing that, though longer, was in warmer climes with the added benefit of a much lower likelihood of inclement sea conditions, a distinct plus before the days of serious ship stabilisation. These assets had potential appeal to premium passengers and, thus, the 'Sunny Southern Route' was coined and promoted by the marketing departments of the three Italian companies.

At the same time, deliveries of new tonnage witnessed the first serious expansion in the number and size of Italian-flag liners for all routes, for the services to South America, South Africa and the Far East besides the North Atlantic run. NGI finally commissioned *Giulio Cesare* and *Duilio*, followed by the larger *Augustus* and *Roma*, the former a motor vessel, the latter a steamship. Lloyd Sabaudo introduced a new *Conte Rosso* (an earlier ship of the name had been contracted from the William Beardmore yard at Dalmuir, Glasgow, but never saw commercial service since she was commandeered while still under construction and completed for the Royal Navy as the aircraft carrier HMS *Argus*) with a sister-ship, *Conte Verde*, and then added *Conte Biancamano* and *Conte Grande*, also of greater tonnage. For Cosulich, its *Saturnia* and *Vulcania* completed a major new building programme amounting to ten significant liners between 1922 and 1928.

By the late 1920s, the Cosulich Line had come under the wing of Lloyd Sabaudo, as a subsidiary to the parent company, leaving the main transatlantic competition between NGI and Lloyd Sabaudo. It was evident that in order to reduce the longer passage time from Genoa, by definition, bigger ships were required. Where a few extra knots could knock hours off the crossing times from the Channel ports, they would potentially take days off the longer route from the Mediterranean. Equally, fully aware of developments in France and Germany, both lines had designs on a generation of much larger ships that could rival

Ile-de-France, *Bremen* and *Europa* for both speed and the grandeur of their appointments. Hence, the NGI planned a ship of around 50,000 gross tons to be named *Rex*, honouring King Vittorio Emanuele III, although it was rumoured that originally the name *Guglielmo Marconi* had been considered. The selected name *Rex* was considered auspicious, recognising the Latin expression *Regis nomen, navis omen*, meaning 'a royal name was a good omen for the future'. Lloyd Sabaudo's challenge took the form of a vessel of slightly lower tonnage and smaller dimensions that eventually emerged as *Conte di Savoia*.

If not the conception, then certainly the realisation of *Rex* and *Conte di Savoia* was intrinsically connected to the economic conditions in Italy at that time along with the policies of the government of Benito Mussolini, elected in October 1922. The new ship plans, which depended on financial backing from the Italian government, evolved against a background in which a policy was adopted for the promotion of greater nationalisation of Italian passenger shipping, the ultimate goal being to merge the companies into strategic groups devoted to zones of regional operation under the full financial control of the state.

Discussions to that end, between the Ministry of Communications, which had responsibility for merchant shipping, and the private operators, commenced in the mid-1920s. Initially, there had been a concept for a new company called Lloyd Littorio that would effectively replace NGI and for which two 40,000 gross ton ships named *Rex* and *Dux* would be built, the latter to be christened in honour of Prime Minister Benito Mussolini known as '*Il Duce*' (The Leader). The construction of both ships would be completely financed by public money. In a sense, where NGI was concerned, this was somewhat academic since the company was already virtually under state control anyway, its finances controlled by the Banca Commerciale Italiana. In the event, though, this proposal was abandoned and, in its place, a pooling agreement between the lines was pursued and ultimately signed in 1928 under which the sailing schedules of NGI and Lloyd Sabaudo would be co-ordinated and a common fare structure implemented. Simultaneous with the pooling agreement, the Italian government created the Istituto di Credito Italiano as a means of financing new ship projects.

Rex under construction on the slipway at Ansaldo, Sestri Ponente, in 1931. (Ansaldo Archives)

Another view of *Rex* on the slipway. Her pronounced counter stern can be clearly seen in this photograph. (Ansaldo Archives)

The launch of *Rex* on 1 August 1931. (Ansaldo Archives)

The two independent concerns, to the extent that they both remained of this status, were reluctant participants in the pooling arrangement but the impact of the Great Depression after October 1929 made closer co-operation imperative. This resulted, ultimately, in the creation of a single combined state-owned holding company, the Italia Flotte Riuniti or Italia Line, from 2 January 1932, the first step in the full nationalisation and reorganisation of Italian shipping.

Meanwhile, as these manoeuvrings were unfolding, the two planned giant liners had been ordered respectively by NGI and Lloyd Sabaudo, and their construction had commenced under the auspices of these, their original, independent owners. But they would never wear the liveries of those companies for the construction contracts were to be assumed by Italia Flotte Riuniti and, on completion, both would enter service sporting the distinctive green, white and red funnel colours of Italia.

The financing arrangements settled, the order for *Rex* was placed on 2 December 1929. In fact, it constituted two elements: the hull construction and propulsion machinery from the Ansaldo SpA shipyard at Sestri Ponente, Genoa, as Yard No. 296, and the fitting out by the Officine Allestimento Riparazione Navali (OARN), the Ship Preparation Workshop. The OARN, which had originated in the mid-nineteenth century at the onset of iron shipbuilding, was another part of the Ansaldo Group of companies, all of which were under the control of the Banca d'Italia.

The formal laying of *Rex*'s first keel plate was performed on 27 April 1930 and the following year, on 1 August, the launching ceremony took place attended by the King and his consort, Queen Elena of Montenegro, who became the ship's patron.

The worsening of the Depression, which led to a 60 per cent reduction in passenger bookings on Italian liners, prompted calls for the construction of *Rex* and *Conte di Savoia* to be terminated, even though it would cost hundreds of shipyard and engine works jobs, for vessels of such size were considered totally inappropriate when the shipping industry was experiencing such a downturn. But the Italian government would have none of it and, on the personal insistence of Mussolini, the order was to press on without delay. He held the view that the two great liners would prove to be prestigious flag wavers for the Italian state and, once they had entered service and international finances had been restored, they would surely pay their way and elevate Italy's standing on the transatlantic route.

For vessels of their size, both ships were built in near record time and some serious mechanical failures encountered when they first commenced operations were attributed to this. On 4 September 1932, just two years and four months after she had been laid down, *Rex* began her preliminary sea trials. Ten days later, during her official speed trials over the measured mile from Punta Chiappa to Punta Mesco, in the Gulf of Genoa, she achieved a speed of 29.2 knots. It would have been higher had one of her turbines not broken down. Nonetheless, she was accepted by her owners when delivered to them at the formal commissioning on 25 September.

Rex had a gross tonnage of 51,062 and her principal dimensions were 880ft length overall and 97ft beam, a measurement that would have permitted passage through the Panama Canal, albeit a transit that was never made. The hull also had slight tumblehome but arising from the fact that two of her upper decks, the saloon deck and the promenade deck, extended out beyond the hull wall by around 3ft on each side, her actual maximum breadth was just under 102ft.

In contrast with the exterior design development introduced by *Bremen* and *Europa*, and even revealed in her consort *Conte di Savoia*, the style of *Rex* was a blend of old and new. For all that, she was an attractive ship with elegant lines and a graceful sheer. Her bow had a raked stem sloping down to a bulbous forefoot but, unlike that of *Bremen* and *Europa*, it did not project forwards but was shaped within the hull, in line with the angle of the stem's incline. Like *Bremen*, *Rex* had forward-facing butts to the hull plating and she also had a streamlined Oertz rudder, design particulars arrived at following extensive model hull tests carried out in a testing tank in Hamburg, Germany. Her hull was constructed with bilge keels to reduce excessive rolling. In technical papers, they were described as being 'fluid dynamic'.

Although cruiser or round sterns had become the vogue, *Rex* retained a counter stern, for which two explanations were given. The first was that it provided an overhang above the caisson gates of the Genoa drydock and the second was that it provided the maximum amount of open space on the aft decks. Topping off her superstructure, she had two squat, slightly inclined funnels that also conveyed a hint of modernity. But her masts, which were unusually tall, in fact the tallest ever fitted to a liner, were a slightly retrograde detraction from a generally sleek overall impression.

Her luxurious interiors – certainly that was the case in first class but better described as 'comfortable' for the lowest echelons – also tended to reflect the traditional fashions of earlier times. There was no art deco modernism as in some of the public spaces of *Conte di Savoia*. Instead, the decorative style of *Rex* harked back to an almost antiquated, classical convention, though it was not as extreme as in earlier NGI liners. The first-class public rooms, imposing and elegant, featured marble columns and ornate murals decorating the ceilings. Described as 'Old World' mode, the rather ostentatious and elaborate style met with some criticism from the trade press but the adopted fashion seems to have met with the approval of her passengers.

In recognition of the selling point of the 'Sunny Southern Route', *Rex* had an open-air lido with swimming pool and verandah, as did *Conte di Savoia*. It is said that this feature introduced the word 'lido' to the American lexicon and from then on the Italia ships were referred to as the 'Lido Liners'. *Rex* also had a spa and gymnasium, both novel amenities.

Rex had ten decks, four of which, decks A to D, ran the full length of the hull. Five of her decks were dedicated, more or less in their entirety, to first class, a disproportionately large allocation to just 25 per cent of the passengers, approaching 50 per cent of the available accommodation space. Her passenger numbers, as built, are recorded as 604 first, 378 special or second, 410 tourist and 866 in third class. The distribution and the overall numbers of berths aboard her later changed, and by the time *Rex* sailed from Genoa on 8 March 1936 her complement

Rex during fitting out. Her twin funnels are in the process of being installed. (Ansaldo Archives)

Completed in September 1932, *Rex* is seen here during her trials. In winter North Atlantic conditions even the hardiest would have been dissuaded from using her exterior lido pool! (Arnold Kludas)

Dressed overall, *Rex* sets off on her maiden voyage. Little was it realised at the time that two days later she would ignominiously break down, taking the gloss from her inaugural sailing. (Italia Line)

Rex seen on her belated maiden arrival at New York following the breakdown off Gibraltar. On a subsequent crossing in August 1933, when she secured the Blue Riband, her landfall was more jubilant. In readiness, she arrived flying a pennant to declare her success. (US Naval Historical & Heritage Command)

had changed to 424 first, 673 tourist and 718 third class. The second or special cabins, which had been treated as interchangeable, had been modified and the majority allocated to tourist class in recognition of the growing tourism demands of middle-income American travellers. The officers and crew numbered 880.

Her propulsion machinery comprised twelve Parsons-type steam turbines, built under licence at the Produzione Sampierdarena engine works, driving quadruple screws through single-reduction gearing. The engine arrangement had each shaft driven by a set of three turbines, one high pressure (HP), one intermediate pressure (IP) and one low pressure (LP), with astern blading on the IP and LP units. Steam at a pressure of 385 psi was supplied by twelve oil-fired water-tube boilers, eight double-ended and four single-ended. With a rated output of 120,000 shp, *Rex* had a service speed of 27 knots, her maximum power output calculated as 142,000 shp. She did not suffer particularly from vibration and she exhibited good sea-keeping qualities.

Rex's maiden voyage to New York commenced from Genoa on 27 September 1932. She called first at Villefranche – Naples was omitted – where she was obliged to anchor because of the absence of quay facilities, before experiencing a severe electrical breakdown off Gibraltar where she arrived two days after first setting out. A major power failure had been caused by an ingress of sea water from a defective boiler into her dynamo chambers, completely disabling two of her three turbo-alternators. Exacerbated by the distribution of the dynamos in the engine spaces, inaccessibility issues added to the problems engineers had to contend with in order to effect repairs. Her Master, Captain Francesco Tarabotto, ordered all electrical services in the crew areas to be switched off in order to preserve whatever power that could still be generated for the benefit of the passengers. Of course, the power interruption affected cooking, cabin air conditioning and telephones, lighting, lifts, water supply and other essential services.

The ship's engineers, under Chief Engineer Luigi Risso, worked tirelessly to restore power, assisted by a number of personnel from the nearby Royal Navy dockyard. Despite this, the hold-up lasted for two days while her fleet mate *Vulcania* was awaited, bringing replacement parts from Genoa. Sadly, before *Rex* was able to resume her voyage on 2 October, around seventy-five disgruntled passengers had aborted their crossing aboard her. Some continued to New York aboard *Vulcania*, while others made their way across Spain and France to join *Europa* then due to sail from Le Havre. All in all, it was a disappointing blemish to *Rex*'s inauguration.

She finally arrived in New York on 7 October 1932, after suffering further difficulties en route, to the usual fanfare and flotilla of small craft that greeted all new ocean liners. New Yorkers endeared themselves to each maiden arrival in their port for it was the focal point of all the foreign shipping efforts designed to forge better links with the New World, and *Rex* was no exception.

Two days after her arrival, *Rex* was opened to the public and a huge number of ex-pat Italians as well as other New York citizens swarmed to Pier 86 to see her. So great were their numbers that the traffic between 42nd and 46th Streets was brought to a standstill and a contingent of more than 250 police officers was despatched to the scene to restore order. Apparently, some 25,000 would-be visitors were turned away, around half of those who had converged on the docks, an indication of the popular reception afforded this new Italian super-liner. Those who did manage to get aboard were charged 25 cents each for the privilege, accruing a handsome sum of more than $5,000.

An equally unruly incident involving *Rex* occurred later, on 12 December 1936, just prior to her eastbound departure that day. Around 5,000 visitors forced the barriers at Pier 59 (a temporary berth while the new Pier 92 was being prepared for Italia) and streamed aboard her. Embarkation had to be suspended while company officials aided by port security officers endeavoured with difficulty to restore order and eject the mob.

Rex finally left on her maiden return crossing twelve days late, on 19 October, having been held up in port waiting for engineers and mechanics to arrive from Italy aboard *Conte Grande* to complete the generator repairs. After making the eastbound crossing, she finally entered Genoa on 26 October to an equally tumultuous welcome. Thereafter, her teething problems resolved, she settled down to a steady and reliable schedule. It is accepted that the British, German and French Lines fared best in the straitened economic conditions of those years but, in fact, the front-line Italia Line ships also achieved respectable results in terms of occupancy levels. *Rex*'s load factor in 1937–38 at 49 per cent compared favourably with the 53 per cent of *Normandie* and the 67 per cent of *Queen Mary*, while that of *Bremen* and *Europa* was at a similar level to *Normandie*. Nonetheless, it was well known that *Rex* and *Conte di Savoia* were not recovering their running costs and, in the name of prestige, they depended on state subsidies to remain viable.

Rex with tugs at Genoa. She is seen flying her Blue Riband pennant. (Authors' collection)

Left: This side view of *Rex* shows off her elegant lines and overall stylish looks. (Authors' collection)

Below: *Rex* moored at the Stazione Marittima, Naples, on the return leg of her record-breaking voyage, with admiring crowds looking on. (Authors' collection)

Although conceived to be able to function as a dual-purpose ship with some cruising in mind, her primary function, as for all the giant super-liners that entered service in the period of the late 1920s and '30s, was for year-round transatlantic passenger service. In the event, *Rex* made only three cruises in total. The first, which commenced from New York on 24 December of her inaugural year took her to Central America and the West Indies and was deemed to have been a great success.

Rex made history on 16 March 1933 when, in mid-ocean, she became the setting of the first ever live radio broadcast made from a ship at sea. The renowned soprano Rosa Ponselle, a regular performer at New York's Metropolitan Opera House, sang Schubert's *Ave Maria*, which was heard by thousands ashore on either side of the Atlantic. At the time, Ponselle was travelling to Italy to fulfil a promise she had made to her mother that, on at least one occasion, she would sing in Italy, the home of opera, at the famous Teatro alla Scala in Milan.

A significant achievement for *Rex* came later that year when she captured the Atlantic Blue Riband from *Europa* for the fastest westbound crossing. The triumphant liner arrived in New York on 16 August 1933 after a passage from Tarifa Point, Spain, to the Ambrose Channel light vessel of four days and fourteen hours duration, at an average speed of 28.92 knots. Her best day's run was 736 miles at an average of 29.61 knots. She became the first record-breaker to fly a blue pennant from her mainmast in recognition of her status and she was also the first to be awarded the Hales Trophy, a laudatory and tangible token honouring her accomplishment, donated by the British MP Harold K. Hales.

On 27 February 1935, *Rex* made her one and only Mediterranean cruise, the itinerary comprising calls at Gibraltar, Cannes, Monaco, Genoa, Naples, Cyprus, Haifa, Port Said and Rhodes. Her third and final cruise, three years later, set out from New York on 29 January 1938. It was deliberately arranged and promoted to steal the thunder from *Normandie*'s Rio cruise of that year. Unlike *Normandie*, *Rex* was able to offer her passengers greater comfort while passing through the tropics as she had been designed for operation in warmer weather conditions. Sponsored by American Express and lasting twenty-five days, the cruise included visits to Cristóbal and Colon on the Panama Canal, Havana, La Guaira, Trinidad and Barbados before *Rex* sailed on to Rio de Janeiro for the annual Mardi Gras.

Mussolini's expansionist foreign adventures in Africa – the invasion of Abyssinia (Ethiopia) in October 1935 and the suppression of colonial resistance in Libya and Somalia – instigated a censure from the League of Nations and the imposition of sanctions. More directly, where *Rex* was concerned, it led the London insurance underwriters to decline to renew cover for her and *Conte di Savoia*, necessitating the Italian state to provide future insurance protection for the two liners.

The first of two detractive experiences for *Rex*, connected with the country's political stance, occurred on 26 November 1936 when she was called upon to participate in a naval review in the Gulf of Genoa in honour of the visiting Admiral Nicholas Horthy, the Hungarian regent, another fascist dictator. Two years later, on 5 May 1938, at an equally injurious event, *Rex* was again part of a naval review parade in the Gulf of Genoa for the visit of Adolf Hitler. As the nation's flagship and a prestigious symbol of the rise of Italy's ranking in the North Atlantic passenger trade, it was natural for a ship of her stature to be present. However, the ignominious sentiment of these displays did little to impress international travellers, notably Americans, who were unsympathetic with the ideals and objectives of the European fascist dictatorships, and this progressively had an impact upon her passenger numbers.

The next phase in the furtherance of the Italian government's policy for the nationalisation of shipping was a convoluted affair, envisaged as the means to improve efficiency by rationalising resources and identifying the old, defunct and obsolete ships for replacement. In part it was also seen as a way to restore the fortunes of the Italian banking system then on the brink of collapse due to the inability of companies, seriously affected by the economic depression, to repay their debts. A work in progress, as mandated by Mussolini, the first move was to create the Istituto per la Ricostruzione Industriale (IRI) in 1933 to take on these debts.

In turn, this organisation created a series of semi-autonomous holding companies described as 'finanziarie' to manage the IRI's state ownership objectives for specific industries. Of these, the Societe Finanziaria Maritima (Finmare) was set up on 21 December 1936 to manage shipping under the broad nationalisation policy. In parallel, also mandated by Mussolini, the Istituto per il Credito Navale was created by Alberto Beneduce, one of the directors of IRI, as a future financing source for shipping, both new construction and operational subsidisation.

Finmare took on the task of organising the national fleet by becoming the parent company of four distinct, self-contained shipping lines, one of which was the Italia Flotte Riuniti, as it was formerly known. Re-registered as the Societa Anonima Italia di Navigazione, it became the owner of the Italia Line from 1 January 1937. Concentrating on the Atlantic routes, north and south, the regenerated concern lost some of its ships and gained others. As part of the reorganisation, some of its ships were switched from the New York route to the Buenos Aires and Valparaiso services. *Rex* and *Conte di Savoia* formed part of

Rex passing at speed, a photograph taken from her consort, *Conte di Savoia*. (Italia Line)

the remodelled fleet, the former transferred to the restyled company for 150 million lire. Other than that, these transactions had few direct consequences for the pair and they continued about their North Atlantic schedules much as before, wearing the same livery.

Later that year, on 24 December, a freak accident occurred while *Rex* was anchoring at Villefranche. It resulted in the death of one crew member and seriously injured two others. Somehow the anchor fell into a shallow area of the anchorage in such a way that the sudden shock caused the anchor chain to break, dragging all the links into the sea. The anchor and chain were recovered by divers on 6 April of the following year. Later, Villefranche was dropped from *Rex*'s port calls, replaced by Cannes where it was still necessary to anchor in the absence of a terminal with berths.

An ominously prescient incident involved *Rex* on 12 May 1938 when she was subjected to a close flyby from three prototype US Army Air Force B-17 Flying Fortress bombers 700 miles off the American coast. The objective of the air force exercise was to demonstrate the ability of the planes to intercept ships at sea but neither the company nor the ship was forewarned of the intended manoeuvre, nor was any prior request made for clearance to carry it out. For the passengers on deck, witnessing the simulated assault by the three aircraft, swooping low over the liner, it was an experience mixed with fascination and consternation. But alarmed by such an unheralded and threatening war game, it triggered a diplomatic fuss as the Italian government lodged a formal protest with its US counterpart. Not remotely thought conceivable at the time, with the benefit of hindsight, it was a strangely prophetic hint of what was to befall *Rex* some six years later.

After the outbreak of the Second World War, *Rex* and *Conte di Savoia*, along with other Italian liners, continued about their routine schedules. On 8 September 1939, *Rex* made her first Genoa to New York crossing wearing neutrality markings on the sides of her hull, recognising that Italy at that time had not entered the conflict as a belligerent. Transatlantic sailings continued through to the spring of 1940 until, lacking certainty as to the invulnerability of *Rex*, her

Arriving at Genoa, *Rex* is assisted to her berth by tugs. (Authors' collection)

A stern port quarter view of *Rex* on 4 February 1938. She had a graceful sheer and looked sleek viewed from every angle. (Authors' collection)

Rex alongside at Rio de Janeiro during her cruise in February 1938. (Italian Liners Association)

crossings were finally suspended. She left New York on 11 May 1940, arriving back at Genoa seven days later at the end of her final eastbound crossing. Little was it realised then but it turned out to be her last ocean voyage altogether.

Rex was laid up at Genoa. Her neutrality signs were removed and she was painted with a wavy white band along her sides at waterline level. The reason for this addition to her colouring is not apparent. Just over two weeks later, on 6 June 1940, *Rex* sailed from Genoa escorted by the naval submarine chaser *Albatros* bound for Venice for lay-up with *Conte di Savoia*. However, en route she was diverted to Pula in Istria, where she was to be employed as an accommodation ship for workers at the Monfalcone shipyard who were engaged refitting the old battleship *Caio Duilio*. For the passage her lights were reduced or shielded and signalling by light between the two ships was not permitted. Arriving at Pula three days later, she was repainted all over in a sand colour and anti-aircraft guns were installed aboard her.

On 15 August 1940, *Rex* was transferred to Trieste, again moved with *Albatros* in attendance. There was no obvious auxiliary employment for her. Consideration was given to adaptation into a troopship but the Italian naval authorities decided she was too large and consequently would be conspicuously vulnerable in that role. Along with *Conte di Savoia*, she was also the subject of studies of suitability for conversion into an aircraft carrier. This, too, was abandoned just as it was later for *Europa*, *Normandie*, and even the Cunard *Queens*, because of structural complications inherent in their accommodation layouts, besides the time and expense that could be involved that would be better spent on more readily attainable naval vessels.

As a result, *Rex* was laid up idle, moored alongside the north side of Pier VI in the Punto Franco Nuovo, where she remained for the next three years.

Everything changed when the Armistice of Cassibile was concluded with Italy and signed on 3 September 1943, announced by the Allies five days later. On that same day, German troops moved to take control of Italy and *Rex* was seized by the invaders the next day, 9 September. She was abandoned to the occupiers, and they set about looting her of equipment and materials, furniture, artwork and ornamental artefacts. Local citizens also removed whatever they could to trade for food, which was desperately scarce.

On 3 March 1944, *Rex*, now a shadow of her former self, was registered with new owners, Mittelmeer Reederei GmbH, Hamburg, a semi-public company set up by the German Reich government on 16 November 1942 to manage shipping in the Mediterranean. Prompted by problems with the supply of men and armaments to German forces in North Africa, the company was authorised to acquire tonnage by expropriation, confiscation or charter. But *Rex* was not in a fit state to go back to sea and there was little prospect, at this stage of the war, of receiving shipyard attention to restore her for any practical purpose. Bearing that out, she remained languishing at her Trieste berth until, on 5 September 1944, it was decided to move *Rex* to an anchorage at Semedela, between Izola and Capo d'Istria. Conflicting views have been expressed as to the reason for this move. The German perspective was that it had become necessary to relocate her to a less-exposed anchorage because of the intensified Allied bombing of Trieste since that June. This suggested a concern for her preservation. The Allies' opinion, on the other hand, was that she was being prepared in readiness to deliberately sink her in order to obstruct the shipping channel into Trieste to deny their use of the port.

Whichever it was, during the tow *Rex* ran aground as the consequence of a manoeuvring error. The following day, 6 September, she was refloated, but having been delayed in her conspicuous predicament she was spotted by Yugoslav partisans, who informed Allied Command of her presence and position. As a result, in the evening two days later the beleaguered liner came under attack by RAF and SAAF (South African Air Force) Bristol Beaufighter light bombers, of the Mediterranean Allied Coastal Command and the Balkan Air Command respectively, from bases at Carnizza (modern-day Karnïza), Slovenia and Biferno.

The sortie reports reveal a clear impression of how the attack unfolded and, strikingly, how rapidly *Rex*, a sitting duck, was fatally overwhelmed. In the first wave of the attack, six aircraft from 39 Squadron, RAF, scored the initial hits on the defenceless ship despite light flak coming from the Trieste harbour area. When the second wave arrived on the scene, six aircraft from 16 Squadron, SAAF, *Rex* was already ablaze and listing to port. Divided into two groups of three, designated 'A' and 'X', the aircraft of this second wing between them hit her repeatedly with numerous rounds of cannon fire and salvos of air-to-ground rocket projectiles. Some of these undershot but possibly struck *Rex* beneath the water; others hit her repeatedly amidships on the still exposed hull and upperworks. It was all over in barely two minutes and in less than an hour and a half the planes were back at their bases.

During the Second World War, *Rex*, like her consort *Conte di Savoia*, was painted in several different colour schemes. This unique and unusual design, whose purpose is unknown, exhibited an undulating light-painted area on her lower hull that extended down to the waterline. (Mario Cicogna)

Rex while under attack from Allied warplanes in September 1944. Some rockets and cannon fire have overshot, but undershot rounds have already breached her hull beneath the water. (RAF/SAAF)

The crippled *Rex*, burning fiercely, rolled over and sank half-submerged in shallow water, her starboard side still exposed from which smoke belched for days. The wrecked ship remained as she lay, partially burnt out, 165 yards offshore in the position 440E, 300N, until after the war had ended. Yet, even then, she became a centre of contention. Under the peace treaty that ended the Second World War in Europe, the territory south of Trieste, where the wreck lay, was controversially ceded to Yugoslavia. The Italian government complained and countless refugees were forced to flee from Istria to get across the new boundary of their homeland, while access to the wreck became a matter of acrimonious dispute.

Denied access to the wreck, the Italians eventually abandoned all hope of recovering her although, realistically, she was probably beyond anything other than a punitively expensive and unjustifiable major rebuild. The Tito regime in Yugoslavia did not want her, had no interest in attempting to raise her, and instead authorised breaking her up where she lay to recover valuable metal. A lengthy process, it took until June 1958 to complete and even then, so it is claimed, some parts of the ship, including one of her screw propellers, part of her double bottom and engine components, still remain on the sea bed off the Slovenian coast. The letters of her name, removed from her hull, and the ship's bell, were saved and have been preserved.

Rex was the third of the great super-liners to fall as a wartime casualty, in many respects bringing the curtain down on what had been a glorious era of glamorous pre-war ocean travel. In her short peacetime career, from September 1932 to May 1940, *Rex* had completed 202 Atlantic crossings (101 round voyages) besides her three cruises. She had carried in excess of 300,000 passengers.

Rex has a mixed legacy. Regarded as having been a symbol of fascist arrogance, she later featured briefly in the film *Amarcord* by Federico Fellini, which mocked the Italian fascist regime that had been responsible for her demise. But she is also remembered as a 'ship of salvation' thanks to the conspicuous kindness and empathy of her crew in assisting some 30,000 Jews to take passage to get to the United States. They followed different escape routes to board her at either Genoa or Cannes, and she became their means of escape from the oppression in Germany and Austria, no doubt sparing them from the concentration camps. Like Germany, Italy had introduced racial laws but a decree was sought and issued that suspended them from being applicable aboard ship.

Rex is also recalled as one of the legendary ships of those times, both for her stylish looks and as Italy's sole super-liner and Atlantic Blue Riband holder.

A sad sight: *Rex* belching smoke, lies completely wrecked on her port side. The attack dealt her a fatal blow and she was left fit only for breaking up. (RAF/SAAF)

FOUR

A SHIP OF COMBINED POWER AND ELEGANCE: THE MIGHTY AND INCOMPARABLE *NORMANDIE*

No radically new large ship appeared on the North Atlantic for nearly ten years following the First World War. Those built in the early 1920s were small and economical, possibly reflecting a conservative stance and belief that the day of the super-liner had passed. The new voyagers of the 1920s had been denied the thrill of crossing in a brand-new liner.

Cunard's big three, *Mauretania*, *Aquitania* and *Berengaria*, were handsome and compatible running consorts but by 1927 they were beginning to show their age. They were all pre-war ships, decorated to the tastes of an era that grew increasingly remote. White Star's trio was also very much last year's model. Into this seeming vacuum, the French Line, more correctly known as the Compagnie Générale Transatlantique, introduced its new ship, *Ile-de-France*, in 1927. The French Line's earlier post-war flagship was *Paris* of 1921, which had set radical trends in its interior design, highlighted in the International Exhibition of Modern Decorative and Industrial Arts in Paris in 1925. *Ile-de-France* took interior design and ship's furnishings a step further in its modern appointments, now known as art deco.

That the passenger shipping scene was on the rise again was confirmed by the entry into service of the German *Bremen* and *Europa*, and the French Line wanted to remain a serious contender at the premier level. A Russian emigré, Vladimir Yourkevitch, who had been a naval architect in Czarist Russia and who had designed battleships for the Russian navy, provided the solution.

Following the Russian Revolution, Yourkevitch had made his way first to Paris. Later, in 1928, he took up an appointment as a naval architect at the Saint-Nazaire shipyard, where he soon rose to prominence. He submitted proposals for the unique hull design of a very large liner for evaluation, a concept that the yard of Chantiers & Ateliers de Penhöet in Saint-Nazaire found convincing. Models were first tested at the Versailles test tank but were later switched to the Hamburg tank, where a 28ft wax model was used. After some 2,000 tests, the design had been refined to its definitive profile, sometimes known as the 'Yourkevitch Principle'.

The French Line floated a shares issue in Britain during 1930 to raise £1,600,000 sterling to help finance the project for a new liner in accordance with the 'Yourkevitch Principle'; a similar issue may have been floated in France. The French government also granted the French Line a loan of 120.5 million francs over twenty years at 5.5 per cent interest towards the cost of building. Thereafter the state would act as guarantor for the interest, the contract between the company and the government stipulating that an annual subsidy of 50 to 150 million francs would also be paid to the French Line commencing in July 1933. The shipbuilders had to build a special berth to accommodate a vessel of such an epic scale. Consequently, Penhöet engineers extended the ways by 300ft under the water from the slipway, a task that took some twenty-one months to complete. The decision to build such a massive ship was not universally popular, some feeling that *Normandie* was a symbol of reckless extravagance in times when the Depression was biting and that the vast sums expended on her could have been put to better use. Against this backdrop the keel to Yard No. T6 was laid on 26 January 1931.

Her construction directly employed some 2,000 shipyard workers while sub-contract work included the supply of Hastie's steering gear from Scotland and the great casting of the 102-ton stern frame from the Skoda works in Czechoslovakia. Also, the four, 16ft diameter, three-bladed propellers were cast in England.

The innovative turbo-electric propulsion system consisted of four two-stage, uni-directional Als-Thom-Zoelly impulse turbines each directly driving a 33,400kW, three-phase Als-Thom alternating current (a.c.) generator that developed 40,000 hp. This in turn, via switchgear and heavy-duty cabling, supplied electrical power to four massive motors that drove the shafts with an output of 160,000 shp. Although each turbo-alternator revolved at 2,430 rev/min, the propulsion motor to which it supplied power operated at 243 rev/min and the four three-bladed propellers, each weighing 23 tons, rotated the same at full speed. A remarkable feature of *Normandie*'s machinery was the choice, unusual for that time, of alternating current. The power supply to each propulsion motor was 5,500–6,000 volts. By using this high-voltage alternating current it was possible to use lighter electric cables than would have been needed by using direct current (d.c.) machines to generate the equivalent power.

In this style of propulsion there was no cumbersome gearing, no long shafting and no astern turbines. When astern manoeuvring was required the polarity of the huge motors was reversed. In addition, very fine tuning of each propeller speed was achieved such that the rotational speed of the screws could be synchronised. Steam was supplied by twenty-nine Sulzer oil-fired, mono-tube, three-drum water-tube boilers at 400 psi and 660°F superheat. These were built by the Penhöet company. There were also four Scotch boilers that supplied saturated steam for auxiliary purposes. The uptakes of the water-tube boilers were divided and taken up the sides of the hull, to be rejoined just below the funnels, as in the fashion of *Vaterland* and *Bismarck*.

Some of the ship's innovative structural features included a clipper bow, rounded and rising up in a pronounced curve. The bow had a bulbous forefoot to add buoyancy in order to minimise water resistance and turbulence and also keep the propellers at their right depth in heavy seas. This was achieved by building the stem from formed plates instead of a conventional straight stem made from cast iron. There was a very distinct sheer rising up to the bow, and above this there was a rounded turtle-backed deck that ran aft for a short way and was then surmounted by an immensely strong V-shaped breakwater designed to prevent heavy seas breaking over the superstructure decks. The ship's rudder weighed 96 tons.

All deck machinery and capstans were concealed beneath the turtle-backed deck. There were four 16-ton anchors, two forward, one emergency in the stem, and one aft. Aft of the breakwater was a small deck extending to the superstructure front, on which was a flush hatch served by two athwartship electric cranes. The superstructure and the bridge fronts were rounded and both the boat deck and bridge were set back from the promenade deck.

Above: Normandie seen on the slipway at the Penhöet shipyard at Saint-Nazaire. (Ch. de l'Atlantique)

Opposite: Launch of *Normandie* on 29 October 1932. (Ch. de l'Atlantique)

Fitting out in the Penhöet Dock on 1 April 1933. (Ch. de l'Atlantique)

Normandie's funnels and masts were installed by July 1934. The aft-most funnel was a dummy. (Ch. de l'Atlantique)

Penhöet Yard No. T6 was launched as *Normandie* by France's First Lady, Madame Lebrun, on 29 October 1932, the event witnessed by 200,000 spectators. As it slid down the ways, the hull displaced 26,830 tons, and by the time *Normandie* was complete some 11 million rivets had been used in her construction. Construction took four years to complete at a cost of £8 million.

Her dimensions were truly epic, surpassing those of *Majestic* by a considerable margin. Originally 79,280 gross tons and 1,029ft length overall, she was the first ship to exceed 1,000ft. Her beam was 119ft and draught 37ft 8in. The height to the top of the foremost funnel was 145ft and the circumference of each funnel was 160ft. All the large funnels were raised on deckhouses, the forward one of the three seated on a deckhouse extending back from the bridge while the aftermost one was surrounded by a fence. Fans and ventilators were placed on top of these deckhouses in a manner that ensured the passenger decks were free of the usual machinery that cluttered the decks of many ships. The funnels were teardrop in section and streamlined, the aft-most being a dummy partly used for dog kennels.

Safety was high on the agenda for the new French ship for, besides carrying thirty lifeboats, *Normandie* had been built with a double bottom, divided into forty compartments running the entire length of the hull. A double hull extended up round the engine and boiler rooms, which were divided into sixty-two small compartments. There was also adequate fire protection, and noise and vibration were greatly attenuated by the use of 'flannel', which was concealed in the walls, decks, joints and throughout the majority of the construction.

Over several hours during her trials in April 1935 she made 31.9 knots. This had been predicted but what had not been anticipated was that *Normandie* hardly consumed more fuel at her service speed of 29 knots than *Ile-de-France*, a vessel of half the size, did at her speed of 23.5 knots, one of the hidden benefits of the Vladimir Yourkevitch hull design. Indeed, this meant the fuel capacity of the ship was such that she carried sufficient to allow her to make the round crossing without refuelling.

Normandie's bow seen when she was high and dry in drydock, showing off the Yourkevitch hull form. (French Line)

The interior décor was a shop window of French artistry created by famous architects and artists of the day who introduced in first class a style embodying the history of Normandy, portrayed on panels and reliefs exhibiting master craftsmanship. Much use was made of decorative glass, plaster of Paris and marble, with lofty public rooms located above and below the promenade deck. The most imposing room on the ship was the first class main dining saloon, 305ft in length, 43ft in width, with a ceiling height of 25ft and a seating capacity for 700. It was panelled entirely in moulded glass tiles with vertical strips of hammered glass. Lighting was by thirty-eight vertical glass brackets, two glass chandeliers and twelve huge glass standards in addition to hidden lamps in a gilded ceiling. At each end there were giant bas-reliefs depicting Normandy's wealth and art. Many other items of artistic décor were displayed in the dining saloon, where exquisite cuisine was offered to the passenger elite of the period.

There were also four private dining rooms, each 17ft by 9ft, entered through eight ornamental bronze doors, their décor exhibiting more mural paintings and bas-reliefs.

The 'deluxe' and first-class suites and staterooms were of the highest standard, and twenty-four of them had a private verandah. The enclosed promenade deck offered a length of 430ft on each side of the ship and had sliding windows in special safety glass with access from the promenade deck via double glazed doors. Other public rooms included the grill room on the boat deck and a modern theatre-cum-cinema that seated 380, shaped to give the best acoustics. There was an 80ft long swimming bath tiled in mosaic and floral-patterned walls, a winter garden and a grand staircase. This latter feature was a magnificent flight of stairs with walls of lacquered gold, and large panels of indirect lighting. The staircase was divided into three sections by rails, and at the head of the staircase was a great bronze statue symbolising Normandy, which led on to a large smoking room. The tourist-class accommodation was aft of first class and consisted of many elegant and lofty public rooms and cabins that, although not featuring the grandeur of first class, offered exceptionally high standards of space and décor, to

Normandie, flying bunting, is seen here ready to leave the shipyard for Le Havre. She was accompanied there by the destroyers *Adroit* and *Foudroyant*. Completion had been achieved after striking workers, anxious about redundancy following her departure, were paid a special bonus payment to resume work. (Ch. de l'Atlantique)

a great extent equivalent to the first-class facilities on an average 20,000 gross ton liner. Third class, at the after end of the ship, was designed with a particular standard of comfort with those passengers in mind, part of it being convertible to tourist class as required.

This was not just a super-liner, it was a 'floating palace' that appealed to both the aristocracy and the proletariat alike! She became dubbed 'the Ship of Light'.

Normandie's original passenger accommodation was for 1,972 passengers, made up of 28 grand luxe, 30 luxe, 790 first class, 16 intermediate class, 645 tourist class, 137 mixed and 350 third class. She also carried a crew of 1,350.

She was scheduled to run on the Le Havre and Southampton to New York route on the westbound run, and the New York to initially Plymouth, later Southampton, and Le Havre route on the homeward run, in consort with *Ile-de-France*, *Paris* and the intermediate liner *De Grasse*. *Normandie*'s call at Southampton was usually by tender when she anchored either off Motherbank in The Solent near Stokes Bay or in Cowes Roads.

She commenced her maiden voyage from Le Havre on 29 May 1935 under the command of Captain René Pugnet. On this, her first westbound crossing she gained the Atlantic Blue Riband with a run from Bishop Rock to Ambrose Light lasting four days three hours and two minutes at an average speed of 29.98 knots, beating the previous record held by Italy's *Rex*. On the return eastbound trip she took four days three hours and twenty-five minutes at an average speed of 30.31 knots, by which she also captured the Blue Riband in that direction from Norddeutscher Lloyd's *Bremen*.

One of the side effects experienced during *Normandie*'s high-speed voyages was unpleasant vibrations, originally thought to be caused by the propellers rotating at around their 240 rev/min. The tourist-class cabins down aft that were located directly over the propellers were vacated and for a while she earned the title of the 'cocktail shaker'. It was said that waiters only half-filled the water glasses to avoid spillage due to the vibrations. One of the advantages of the turbo-electric drive was the sensitive control that permitted the incremental easing back of the propeller revolutions until the vibration ceased, potentially alleviating the discomfort. It also helped to identify the speed at which vibration manifested itself. However, this would have resulted in a loss of speed and a delay in the liner's operational schedule.

A couple of months after *Normandie*'s maiden voyage, on 5 July 1935, Captain Pugnet's wife died following a long illness just as his ship was leaving Le Havre. Captain Pugnet would retire after thirty-nine years at sea in July 1936. He had reached the applicable age limit but his wife's death may have hastened his decision. He was succeeded by Captain Pierre Thoreux. In August 1935, *Normandie*, with 1,800 tourists on board, anchored in Cowes Roads and the ship was opened for inspection.

In October 1935 it was decided that she would lay up for the coming winter. Notwithstanding her short period in service, the French Line reported that she had made a profit of around 15 million francs (£200,000) during her first season. On 23 October, just prior to *Normandie* sailing from New York on her last eastbound voyage of the season, the Duke of Sutherland, the Chairman of the Hales Trophy International Committee, presented Captain René Pugnet with the Hales Trophy for capturing the Blue Riband. This was in the presence of the donor, Harold Keates Hales, MP, and other members of the committee. The members of the International Committee had been guests of the French Line for the round voyage.

During her winter refit from 1 November 1935 to 25 May 1936, alterations were made to *Normandie*'s accommodation. Strengthening members in the form of 'brake stanchions' (a form of shock absorber) were introduced from top to bottom into the aft quarter compartments and the first-class terrace was replaced by a new tourist-class lounge. In addition, her thrust blocks aft of the main drive motors were widened and stiffened; also, the shaft brackets were modified with a view to reducing water turbulence near the propellers. Tests in the Hamburg tank had indicated that vibration could be reduced if the outer screws were transposed and the polarity of the motors reversed. This configuration had worked satisfactorily on *Bremen*. The advice was followed and the change made. The turbine governors were fitted with a differential mechanism so that the ship's propellers could be better synchronised. This was done in an attempt to stiffen up the hull to absorb any future vibration. During this time her gross tonnage was increased to 82,779, then again later to 83,423 after another refit.

A trials view of *Normandie*, a true thoroughbred. Since 1913, German-built super-liners had dominated as the world's largest passenger vessels but *Normandie*'s massive size toppled their heavyweight crowns. (French Line)

Normandie making her maiden arrival at New York on 3 June 1935 having just secured the Atlantic Blue Riband westbound. On her return voyage she took the eastbound record. (US Naval Historical & Heritage Command NH93743)

Normandie left Le Havre on 15 November 1935 with a full technical staff, including vibrational analysis specialists, from where she sailed for Saint-Nazaire to pick up more builder's engineers before heading off into the Bay of Biscay for three days of trials in very bad weather conditions. A day later, while proceeding at full speed, it was found that by the reversal of the outer screws the vibration had decreased by 20 per cent. However, areas in way of the brake stanchions still registered vibration no matter how much the engineers tried to rejig them. The propeller synchronisation also caused additional vibration. Another measure that was attempted and implemented was the cutting of an expansion joint in the superstructure just aft of the Cafe Grill.

The most promising solution was the fitting of a set of four-bladed propellers cast by Du Barre. These had not been ready at the time *Normandie* came in for her refit but by April they had been fitted and she was ready for further trials prior to returning to service. On 28 April 1936 she left Le Havre under the command of Captain Pierre Thoreux for trials off Ushant. On these trials she achieved 30 knots without any significant discernible vibration and the positive results proved beyond expectation.

Normandie returned to Le Havre on 30 April and was scheduled to sail on the following day on the first voyage of her second season, with the elimination of the vibration problem having been reported to the French press. Apparently just after the press conference, the ship's Chief Engineer, Jean Hazard, alerted the Captain to report that a diver had been sent down to inspect the hull and had stated that only three screws were in good shape! There followed a slight panic and a company diver was sent down to confirm these findings and upon his return he confirmed that the port inner screw was indeed missing, an extremely rare occurrence.

In order to meet her schedule, *Normandie* was hastily drydocked and, overnight during a double shift of workmen, both inner shafts were refitted with the original three-bladed propellers. She was eventually ready for service and left Le Havre on 5 May with 629 passengers for her first voyage of the season, beating the brand-new *Queen Mary*'s entry into service by three weeks. But the vibration problem returned! Her average speed for the subsequent, second round voyage was 23.31 knots, which caused disappointment among some passengers who believed she was vibration free.

One might wonder how the loss of the aforementioned screw had not been picked up earlier during the trials. A vessel in a heavy seaway or a ship that has shed a screw experiences 'racing' with no load on the shaft in the water. This normally applies to diesel or turbine-driven ships. It's an old marine engineering conundrum: 'does a turbo-electric-driven ship experience racing when the propeller comes out of the water?' In theory, the shaft should rotate at the speed of the electric motor but, apparently, it all depends on the sensitiveness of the turbine governing. The motor cannot race on its own. Its speed is absolutely tied up with that of the alternator, which in turn is coupled to the speed of the turbine to which it is directly driven. This points to the shedding probably having taken place at much lower propeller revolutions or during manoeuvring, when any racing may not have been significantly detectable. No doubt, it was lost in the tidal basin at Le Havre. By the time of the *Normandie*'s return to Le Havre on 16 May, a replacement four-bladed propeller had been cast and made ready in record time and it was fitted before she departed for New York four days later.

While *Normandie* was anchored at the Motherbank, for Southampton, on 22 June 1936, in the process of disembarking passengers from America by tender, five Blackburn Baffin torpedo bombers of 'A' Flight (Torpedo Training) from RAF Gosport were practising over The Solent. The leader decided they should fly around *Normandie* for a closer look. One of the planes piloted by Flying Officer Guy Kennedy Horsey, who also held the rank of lieutenant RN, flew too close and his plane's port wing struck the crane on the liner's foredeck, snapping it off, before somersaulting and collapsing onto the ship's bow in a wrecked state. Luckily, the pilot was unhurt and later taken off by tender. The wreckage of the aircraft had to remain on board *Normandie* as it was necessary for

Normandie approaching her berth at New York's Pier 88. The luxury, the décor, the cuisine, together made for a memorable experience, primarily focused on her wealthier travellers. (Underwood Archives, Inc. / Alamy Stock Photo)

Normandie at anchor, tendering in The Solent. (Peter Roberts)

Another view of *Normandie* tendering at Spithead with the Red Funnel tender *Calshot* and two of the company's paddle steamers along her port side. Note the cars awaiting off-shipment on the starboard side. (David Hutchings collection)

her to sail due to the tide. Thus, it was conveyed to Le Havre, from where an RAF salvage team later retrieved the wreckage.

The damaged crane had been in the process of hoisting a limousine, belonging to the MP Arthur Evans, onto a tender. Likewise, the vehicle was deposited heavily onto the foredeck, precariously lodged against the ship's deck edge, but fortunately it did not fall onto the tender below or into the sea. This, too, had to go to France with the ship for offloading. Somewhat ironically, *Normandie* herself was actually considered for possible conversion into an aircraft carrier some seven or eight years later, so a carrier-borne plane pranging on her deck may have been in some way prophetic (see page 98).

Following *Queen Mary*'s entry into service in May 1936 and her subsequent capture of the Blue Riband, the French Line decided to go for an increase in speed and commissioned a third set of 16ft 7in diameter, four-bladed propellers, which were fitted during the winter of 1936–37. In addition, the main generating alternators and drive motors were completely overhauled and the supplementary steam nozzles on the turbines opened to increase the steam flow. This latter modification increased her power output to 180,000 shp, which, at a propeller speed of 231 rev/min, produced a ship speed of 32 knots. The first-class spaces were redesignated as the new cabin class and second class became tourist class in keeping with the changing trends of the time.

During the following year, 1937, she made thirty-six Atlantic crossings in all weathers, and her average speed worked out at 28.87 knots. From 18 to 22 March that year she regained the speed title from *Queen Mary* with a time of four days and six minutes at an average speed of 30.99 on the eastern passage. A few months later, between 29 July and 8 August, she bettered her times with a westbound crossing of three days twenty-three hours and two minutes at 30.58 knots, and eastbound at three days twenty-two hours and seven minutes at 31.2 knots. *Normandie* had given everything she had! Thanks to the modifications, she had achieved a maximum power output of 198,850 shp, consuming only a ton of fuel oil every minute.

The aftermath of the aircraft crash onto *Normandie*'s foredeck. The errant aircraft, which lies mangled on her prow, had to be taken to Le Havre for removal. It is a supposition but probably the limousine, precariously balanced on the right side of the foredeck, also had to be taken to France aboard *Normandie*. (Authors' collection)

Normandie seen in Guanabara Bay, on the occasion of her first cruise to Rio de Janeiro in February 1938. (Authors' collection)

Normandie in the Joubert Lock at the Chantiers Penhöet, Saint-Nazaire. (Authors' collection)

This was to be her last record-breaking effort before she conceded in both directions to *Queen Mary* the following year.

One of *Normandie*'s design features was a 'high' transverse metacentric height (also known as GM to naval architects, where G is the centre of gravity of the ship). This meant that during a heavy roll in a rough seaway, *Normandie* would restore herself to the upright swiftly. This was said to be a 'stiff' condition and she was known to be a 'stiff, snappy roller'. The downside of this invisible dimension (10ft on *Queen Mary 2* and 5ft on *Queen Elizabeth 2*) meant that the vessel would sometimes take a heavy pounding in very rough weather.

Arriving at Southampton from New York on 6 September 1937 were the actor Douglas Fairbanks Sr and his wife, the former Lady Sylvia Ashley. On the voyage Mr Fairbanks, who prided himself by keeping fit for action in his swashbuckling roles, would exercise by running around the clear decks early in the morning to avoid his routine from being disturbed by any passengers milling about. Later that same month, another of *Normandie*'s inboard propellers was

Normandie at speed, a study of power and elegance. Her last Atlantic crossing concluded with her arrival at Pier 88, New York, on 28 August 1939, ending her short line-service career. (French Line)

shed but she was able to steam the round transatlantic voyage on her three remaining screws.

On 5 February 1938, *Normandie* undertook her first off-season twenty-four-night cruise to South America, sailing from New York with 975 passengers, all receiving first-class service. Although it was considered a record list for such a voyage, ironically the ship had accommodation for 1,972 passengers, which meant that realistically 1,000 berths were left unfilled. The itinerary was quite unique: New York to Nassau, Trinidad and Rio de Janeiro outwards, arriving on 15 February with four nights in Rio. The return leg was via Barbados and Martinique, arriving back in New York on 27 February. It was deemed such a financial and social success that the same cruise itinerary was repeated in 1939.

During April 1938, after *Normandie*'s Rio cruise, her propellers were changed once again. It is for consideration whether the scheduled change of propellers was to regain the speed record for prestigious purposes or for routine maintenance due to constant erosion. It seems through the ship's short career she rarely enjoyed full occupancy but size and speed were much desired. By July 1938 *Normandie* had made her 100th Atlantic crossing. In the three years of service, she had covered some 334,000 nautical miles at an average speed of 28.54 knots. Along with *Ile-de-France*, *Normandie* maintained her high-speed super-luxury service between Le Havre and New York up until the Second World War.

In the early part of 1939, work commenced on a French comedy film *Paris-New York*. Its plot had a group of *Normandie*'s passengers headed from France to see the 1939 New York City World's Fair. These included two crooks, an American banker and his daughter and two police officers. En route a valuable diamond goes missing. The passengers were played by the top French actors of the late 1930s; Gaby Morlay, Michel Simon, Claude Dauphin and Jules Berg. It was supposed to be a comedy with many sub plots, and it was shot at the Cité Elgé Studios in Paris as well as on board *Normandie*. The film's sets were designed by Andrej Andrejou. It was produced during the 'phoney war' period and released shortly before the fall of France in June 1940. It is hard to comprehend what audiences felt when the film came out as *Normandie* was interned in New York away from her home port and was never to return again.

Early in April 1939, following her 1939 Rio cruise, *Normandie* entered the drydock at Le Havre for routine maintenance and inspection. While there, *Paris*, which was berthed at the nearby quay, caught fire during the night of 19 April, a blaze that spread steadily until the interior of the ship was gutted. To extinguish the fire, she was swamped with water by Le Havre firemen. The following day *Paris* heeled over away from her berth and came to rest on her side, blocking the entrance to the drydock in which *Normandie* lay. Her masts and funnels had to be cut away before *Normandie* could be released.

With the rise of German territorial expansion and the impending war gathering, *Normandie* departed Le Havre on 22 August and arrived in New York on 28 August 1939. She was ordered by the company to stay there and remain 'laid up in safety' alongside the French Line terminal at Pier 88 in Manhattan. In all she had completed 277 voyages, 139 westbound and 138 eastbound, but was fated never to sail again. There she stayed inactive, in comparative safety, for over two years until after the attack on Pearl Harbor on 7 December 1941 that precipitated America's entry into the war.

On 7 March 1940, when *Queen Elizabeth* arrived from Glasgow the adjacent berths to *Normandie* were occupied by Cunard's *Queen Mary* and *Queen Elizabeth* in their wartime grey livery and this was the only time that the three largest liners in the world were moored together.

Five days later she was confiscated by the US Maritime Commission with the view of using her as a much-needed troop carrier for the conflict against the Axis powers. On 24 December she was requisitioned by the US Navy and renamed USS *Lafayette* (AP 53). As such it was planned to start her conversion to a troopship with immediate effect, with a target completion date of 1 February 1942. In the past such a mammoth conversion task would normally have been carried out in a Naval Dockyard but in this instance, this work of national importance was contracted out to a private shipyard called Robins Dry Dock & Repair Co. As such the conversion work was carried out at Pier 88, Manhattan.

The conversion had to comply with unduly complicated US Navy requirements and at times some useless specifications, which made it impossible to complete within the scheduled two months allocated. This was, it is claimed, a melting pot of labour troubles, politics and in some places lack of supervision. The consequential melee of desperate US Navy ambitions balanced against civilian apathy caused hazardous conditions that would have been avoided under less-hurried circumstances. Sometimes, civilian shipyard workers overseen by US Navy overseers did not make for a successful mix!

The US authorities treated *Normandie*'s crew with contempt, even though they had stood by on a care and maintenance basis while the ship was in New York. It was unlike the conversion of *Vaterland* to *Leviathan* as here the necessary expertise of persons who were familiar with the ship's systems was on hand, but it was ignored.

Everything that could be was removed from the ship: fittings, fixtures, valuable works of art, statues, crockery, flatware and 18,000 bottles of expensive wine. Some 2,400 trucks were used to transport all these costly interiors at a cost of $100,000 and several local warehouses were rented for their storage. French operational

On fire at New York as USS *Lafayette* on 9 February 1942. Her already-painted camouflage design can be seen clearly. By this time she had a 20-degree list to port. (US National Archives 80-G-434351)

In her death throes, the capsized wreck of USS *Lafayette* on 10 February 1942. Firefighting continues, but it is in vain. (US National Archives)

manuals had to be translated, even though members of *Normandie*'s crew were on hand to explain the contents in English.

Both internal and external work began apace, while *Lafayette* was also being dazzle painted in her wartime livery. The integral firefighting system was overridden and electrical connections to the fire doors were disconnected. Out of a total of 666 fire extinguishers, only ten had been converted to meet American specifications and more than 50 per cent were in either a useless condition or only partially filled. Fire-watching procedure was not up to strength. Into this disorganised scenario some 11,000 oil-impregnated kapok

The capsized wreck of *Lafayette* dominated by her three distinctive funnels. They were removed during the subsequent salvage operation. She had been due to sail to Boston under the command of Captain Robert C. Coman, USN, on 14 February 1942, just five days after the fire occurred. (US National Archives)

life preservers were stowed on board temporarily in the grand salon. All because the conversion work was running behind schedule.

On 9 February 1942, in this same salon, civilian workmen were removing a quartet of tall lamp standards by flame-cutting their skeletal, heavy steel tripods. White hot steel droplets cascaded down from the last of these features to be removed onto the stack of life preservers, setting them alight. By 2.37 p.m. the flames were taking hold, fanned by the wind blowing down the Hudson River and the ship's ventilation system helping the fire to spread rapidly. Those on the bridge were unable to trigger the fire alarms because no one was familiar with the bridge systems layout and, for the same reason, the heating and ventilation system could not be turned off. Firefighting equipment and procedures were ineffectual and with smoke filling the engine room through forced ventilation, the auxiliary boiler steam furnaces were shut down and *Lafayette* became a dead ship. Chaos reigned. More importantly, who was in charge?

Fortunately, at 2.45 p.m. the New York City Fire Department (NYCFD) were alerted and scrambled. The firefighting was left to the NYCFD, whose firefighters came aboard and in short order ten fire tugs were soon alongside spraying tons of water onto her superstructure. The fire department eventually extinguished the flames with thousands of gallons of water that accumulated in the upper parts of the ship from where it could not escape, partly because the scuppers were overwhelmed. With water trapped on her upper levels, she became top heavy.

Somewhere in the cavernous dining room was the unseen vertical distance connecting two points in space. Had it been plotted, it would have indicated a small, though positive metacentric height (see page 93) but for each gallon poured aboard, that slender margin of stability was reduced. *Lafayette* took on a list to port and began to heel away from the pier. This was arrested temporarily when the low tide allowed the ship to settle on an extended slip.

Trying to navigate the various layers of security to Pier 88 was Vladimir Yourkevitch, the only man with the invaluable knowledge of where the ship's seacocks could be found even in the lifeless black engine room. If these were opened the ship could settle on an even keel. Admiral Adolphus Andrews, the US Navy Port Director, refused Yourkevitch entry and his advice, telling him to mind his own business with the remark, 'This is a Navy job!'

Early the next morning of 10 February, as the tide began to flow, the great ship heeled over onto her port side alongside Pier 88. She had passed from passenger super-liner to naval auxiliary to hulk within the space of seven weeks. Thus, the once proud and mighty 'Ship of Light' had transmogrified into the status of a vast beached whale. It was a similar fate to that which had befallen *Normandie*'s consort *Paris* while on fire at Le Havre during April 1939, related earlier.

Two and a half months after the fire, the mammoth task of salvaging her commenced. It was one of the greatest feats of salvage at the time as there was no previous experience in raising vessels of this size available to guide her salvors. The Navy Department's Bureau of Ships contracted Merritt Chapman & Scott, international salvage experts, under Captain John Tooker, to raise the capsized ship under the oversight of Commander William Sullivan.

This was accomplished by following a four-phase approach:

[i] Removal of the ship's superstructure and top hamper down to the main deck in the hull.

[ii] Clearance of all debris and wreckage surrounding the ship which obstructed areas in way of Piers 88 and 90.

[iii] Erection of bulkheads at strategic points within the sunken hull to effect a series of watertight compartments. When the *Lafayette* had turned over onto her beam ends, only a small number of her watertight doors had been closed while many apertures on her port side, the side resting on the bottom, had been left open.

[iv] Pump the sealed watertight compartments dry. In this way the ship's inherent stability would right her.

One of the most innovative pieces of equipment used to plug the portholes was called a 'Tooker Patch' (named after the Salvage Master). Made of a 3in-thick piece of wood cut in the shape of a porthole, it resembled a boiler manhole door. It was hinged centrally such that the diver could insert it through a porthole from the outside, then open it out and secure it by a strong back. The skill was in the hands of the divers, who had to feel their way along the hull in black muddy conditions to fit each one. Sometimes, it took an entire week to install just one through the effort involved and the appalling conditions of mud, filth and darkness within which they worked. Nonetheless, in all some 356 Tooker Patches were fitted.

By May 1942 a veritable shanty town of salvage experts and engineers was surrounding the hull. There were numerous access holes, cargo hatches, open doors and portholes, in fact some 4,500 openings, all of which needed to be sealed. Meanwhile, scaffolding was erected over the exposed hull for future stages of the operation. Installation of ninety-eight pumps commenced in November 1942 each capable of pumping a total of 2.24 million gallons (10,000 tons) of water per hour. They were located such that each was on a hinged platform towards the ship's upper decks. In this manner they could be adjusted relative to the ship as it gradually returned to an even keel. Pumping started on 4 August 1943 and the long process progressed slowly with the pumps going flat out to compensate for inaccessible areas having unsealed apertures.

As the giant hulk regained its buoyancy incrementally, on the night of 15–16 September 1943 the *Lafayette* finally floated independently for the first time since her capsize. The *Lafayette* was the largest vessel ever to be refloated and the salvage operation was also the most expensive at a cost of $18 million. With her watertight integrity sufficiently restored, she was handed back to the US Navy on 27 October 1943, who redesignated her as potential transport and aircraft carrier (AP-4) and on 3 November 1943 she was towed to the navy drydock at Bayonne New Jersey for further inspection and assessment.

The report of the House of Representatives Committee on Naval Affairs' investigation into the fire and capsizal of USS *Lafayette* found the causes to have been:

1. Carelessness in the manner in which the stanchions were burnt
2. Absence of co-ordination between the activities of the various units on board
3. Divided authority and lack of a unified command
4. The undue haste, indecision and lack of careful planning in connection with the conversion

There was no apportionment of individual blame.

That said, it could be argued that it was the failure of senior naval personnel to engage with Vladimir Yourkevitch, the ship's designer, and act upon his specialist advice which led to her capsizing on her side. The consequential outcomes emanating from that failure were major salvage expense, the loss of a valuable asset for the ensuing war and the lengthy obstruction of a port facility, besides the prevention of this wonder ship from ever returning to commercial service after the war.

As she had been originally designed as a passenger liner, it was felt that she would prove unsuitable as an aircraft carrier, quite apart from which there had been enough money, manpower and resources diverted away from the war effort to the *Normandie* conversion and later her salvage. After languishing, laid up for twenty months, she was stricken from the US Navy register on 11 October 1945 and transferred back to the US Maritime Commission for disposal. In all this time her designer, Vladimir Yourkevitch, still held the hope that she could be rebuilt as a smaller liner but his plans came to nought: on 3 October 1946 the hull was sold to Lipsett Incorporated (Ship Breakers) for $161,180 (a tiny fraction of the salvage cost) and she was broken up.

★ ★ ★

Looking back at *Normandie*, a stark and irrefutable statistic emerges. As the super-liner, at 83,423 gross tons she was the largest merchant ship sunk in both world wars. As USS *Lafayette*, at 66,400 tons displacement, she was the largest 'warship' sunk in both world wars – all by Allied action!

FIVE

TRANSATLANTIC PINNACLE: CUNARD'S WEEKLY EXPRESS SERVICE DUO, *QUEEN MARY* AND *QUEEN ELIZABETH*

The decade of the 1920s had been in all respects a major turning point for the passenger services across the North Atlantic. Still reeling from the impact on trade of the recent war, there followed a period of recession exacerbated by the introduction of laws severely restricting immigration into the United States, a vital source of revenue to the shipping lines seeking to recover from the wartime impediments.

It had been thought that there was no future for new giant liners like the former *Imperator* trio following the cuts to United States immigration but, with recognition that the lost immigrant traffic could be substituted with the materialising trade from American tourists, a revitalised demand for passage spaces, followed by the inauguration of the French Line's 43,000 gross ton *Ile-de-France* in 1927, induced a new race for size and speed.

In quick succession, a reinvigorated Germany had commissioned the stylish *Bremen* and *Europa*, their streamlined exterior lines complementing their modernistic interiors, reinforcing the design transition to a new era. But the crash on the Wall Street Stock Exchange between 24 and 29 October 1929 should have been the writing on the wall. Predicting the future, particularly of financial affairs, is not an exact science, however, and initially the Great Depression that followed did not dampen the aspirations for more giant vessels for the North Atlantic passenger trade.

Cunard was facing unprecedented competition on the North Atlantic run from continental operators, challenging the British company on the route it had dominated for decades. However, Cunard had no intention of competing with *Ile-de-France*, *Bremen* and *Europa* on a like-for-like basis. Before any of these three liners had entered service, Cunard, under the foresight of Sir Percy Bates who became chairman from 1930, was about to up the ante by pursuing the goal of a regular two-ship, four-and-a-half-day weekly express service from either side of the Atlantic. Departures would take place each Wednesday afternoon from Southampton and New York with arrivals on the following Monday morning, rotating after a two-day turnround. The brainchild of Sir Percy Bates, it was referred to as the 'Big Ship' policy. It would represent the zenith of the transatlantic passenger trade in terms of size, speed and regularity.

It was calculated that to achieve this schedule a service speed of 29 knots was required, which necessitated massive steam turbine engines, the largest then contemplated, able to generate 200,000 shp. Moreover, because of the hull space such engines would occupy, along with the need to accommodate some 2,100 passengers in order to make the operation viable, plus 1,100 crew to tend to their every need, ships of much greater tonnage and dimensions would be essential. The specification that emerged from these calculations was for two 80,000 gross ton ships, an increase in tonnage terms at that time of 40 per cent, with a length of more than 1,000ft.

As early as 1926, these outline plans were presented for consideration to John Brown & Company, shipbuilders at Clydebank, the yard that had been Cunard's preferred builder for more than a quarter of a century. From these initial steps, the 'Big Ship' project unfolded and, despite the decline in passenger numbers in the wake of the Wall Street Crash and the ensuing Great Depression, Cunard felt comfortable to proceed with the first of the two ships at its own expense, requiring no government loans or subsidies. Thus No. 534, as she was designated, was ordered from John Brown's on 1 December 1930, her keel laid on 31 January 1931 and entry into service planned for 1933. Together with her consort, they would replace the ageing *Mauretania*, *Aquitania* and *Berengaria* – two ships in place of three for the line's premier service.

In the event though, commencement of *Queen Mary*'s construction, the first of the pair, could not have been more inauspiciously timed, for as the financial world was plunged into a spiralling collapse it became evident that Cunard's optimistic intentions of drawing off its own capital and commercial funds could not weather the economic storm. With passenger bookings down by a third, and falling, the impact on revenues began to bite and the company's liquidity was threatened. As 1931 unfolded the company was struggling to keep its existing fleet operational, let alone to continue to finance the construction of No. 534. Sailings of *Berengaria* and *Aquitania* were reduced and *Mauretania*, painted white, was sent cruising. Even then, further curtailment was called for and the Cunard board had no choice but to make a cataclysmic decision. On 11 December 1931 the company announced that the construction of the new ship was to be suspended. Immediately, an air of gloom descended upon Clydebank as thousands of shipyard workers were laid off, confronting destitution for themselves and their families, and the idle monster, resting incomplete on the slipway, soon became a symbol of the blight afflicting the entire nation. But *Queen Mary* was not abandoned. Her order was not cancelled. It was said that her planned launch for May 1932 was only postponed, although many feared it would be cancelled altogether.

Queen Mary remained idle on the stocks for the next twenty-seven months, while not a sound emanated from the deserted shipyard. The future looked bleak indeed, a far cry from the glorious inauguration of a new era on the Atlantic that had been anticipated.

No. 534, her identity before she became *Queen Mary*, on the slipway at John Brown's in 1931, prior to the suspension of construction. (Glasgow University Archives)

Left: Launch day, 26 September 1934. Unusually, *Queen Mary* was launched with her hull painted white. (Sueddeutsche Zeitung Photo / Alamy Stock Photo)

Below left: Fitting out at Clydebank, *Queen Mary* nears completion. (Pamlin)

Below right: Queen Mary is nursed along the River Clyde for her trials and delivery. She would never return to Clydebank. (Authors' collection)

During her trials at full speed off Arran in the Firth of Clyde, *Queen Mary* exuded an impression of power and majesty. (Glasgow University Archives)

Ironically, during the year in which *Queen Mary* had been started and then abruptly halted, Cunard had entered into negotiations with the Royal Mail Group with a view to buying out the White Star Line, its principal British rival, then owned by Royal Mail. Nothing came of the endeavour but when the government agreed to extend to Cunard a loan of £9.5 million to permit resumption of construction of its super-liner and towards the building of the second ship, a condition of the deal was that the Cunard and White Star companies should merge. White Star had been contemplating a giant 60,000 gross ton ship of its own but the government was not prepared to underwrite grand plans emanating from both concerns, especially as the vessels would become competitors. So it was that the Cunard White Star Line came into being on 1 January 1934, Cunard holding 62 per cent of the equity, White Star the balance.

Work on No. 534 resumed on 3 April 1934, a jubilant day not only for the John Brown shipyard employees, led back into the yard by a band of pipers, but for Scotland and the UK as a whole. The ship was well behind schedule and would now face a new rival in the form of the French Line's *Normandie*, of comparable size and performance, which was close to completion. Her construction had not been interrupted as had happened to the British ship, and she was now destined to enter service a year ahead of *Queen Mary*.

Spurred on by their gratitude at being employed again, the Clydeside workforce set about completing the giant Cunarder with zeal and she was ready for launching on 26 September 1934, christened by Her Majesty Queen Mary. The story of how she came to have her name, arising from a mix-up in royal protocol with King George V, may be apocryphal or true. It has been repeated many times. Briefly, while the name for the new ship was under consideration, it is claimed that His Majesty King George V was approached by Cunard for approval to name her after Britain's greatest queen, referring to Victoria. The King took it as a compliment to his wife and consort, and agreed to the name 'Queen Mary' on that understanding. This may well be an urban myth, but it provides an interesting explanation as to why Cunard deviated from its long-held tradition of bestowing names on its liners with the suffix 'ia', for example *Umbria* and *Etruria*, *Lucania* and *Campania*, *Lusitania* and *Mauretania* etc. However this nomenclature change was arrived at, instead of being named *Victoria*, as it seems had been both intended and anticipated, No. 534 became *Queen Mary*, launching a unique pedigree that has subsequently embraced six more great passenger ships, three of them super-liners. Alternatively, the name may have been decided, it is suggested, as a diplomatic gesture to the White Star element of the combined company, avoiding any name that ended in either 'ia' or 'ic'.

Queen Mary left Clydebank for the commencement of her sea trials in April 1936. She was nursed along the River Clyde on a one-way trip, for she would never again return to her birthplace. Over the measured mile off the Isle of Arran in the Firth of Clyde, she achieved a top speed of 34 knots and it was evident that she had the capability to take the Atlantic Blue Riband from *Normandie*, which had herself taken the westbound record from *Rex* and the eastbound record from *Bremen* the previous May.

Approaching Manhattan, *Queen Mary* arrives at the end of a westbound Atlantic crossing. (Authors' collection)

The original conception of *Queen Mary* had been for a larger *Aquitania*, with three funnels, one deck fewer and a counter stern, but her looks as completed were improved by a cruiser stern, although even then her design remained less modern than her contemporaries. Instead, bulldog British, she looked stalwart and powerful. A true thoroughbred that could punch its way through the most mountainous seas.

As completed, *Queen Mary* measured 80,774 gross tons. Her overall length was 1,019ft and although, by that measure, she was shorter than *Normandie* by 10ft, her waterline length was greater by 13ft. *Queen Mary* had a beam of 119ft, a moulded depth of 118ft, a draught of 39ft and twelve decks. She had cost £3.1 million to build.

When she entered service, passenger grading designations for the transatlantic passenger liners had been amended following an acrimonious dispute among the North Atlantic Conference companies. As a result, *Queen Mary*'s initial passenger breakdown was 776 cabin class (the top grade), 784 tourist class and 579 third class. Her crew numbered 1,101, making a total complement of 3,240 persons and giving her a crew to passenger ratio of 1:1.95.

Befitting her status as the new Cunard flagship, *Queen Mary* was renowned for her spectacular and elegant, if somewhat restrained, interior appointments and decorations, being very much in keeping with the 'look' of the 1930s but projecting more of a British take on the art deco style, commonly referred to as 'Odeon'. It was a tasteful blend of various contemporary design elements notable for the extensive use of polished exotic woods and veneers.

She was powered by four sets of four Parsons impulse-reaction, single-reduction steam turbines geared to quadruple screws. Steam at a pressure of 400 psi and superheated to 700°F was supplied by twenty-four water-tube boilers constructed under licence from Yarrow & Co., each fitted with seven oil burners and for which the ship was capable of producing 300 tons of softened feed water daily. The engines were distributed along the length of her hull, forward to aft, five boiler rooms, two as pairs and one single, separated by generator spaces and followed by two turbine rooms. The maximum power output of 212,000 shp provided for a top speed of 33 knots and of 160,000 shp for a service speed of 28.5 knots. In total, the power and propulsion machinery occupied 30 per cent of the hull capacity.

Despite all the protestations that there was no interest on Cunard's part in entering a competition for speed honours, there was a great sense of patriotic rivalry between the French and British populations for their respective flagships. Arising from the expectation of record speeds amid this fervent public interest, the competition between these two great super-liners, officially endorsed or otherwise, became a feature of the transatlantic scene through to the end of the 1930s. It became the 'space race' of its day.

Queen Mary commenced her maiden voyage from Southampton on 27 May 1936 among impassioned public anticipation that she would break the speed record on her inaugural crossing and wrest the Atlantic Blue Riband from her Gallic opponent. But it was not to be, for her progress was hampered, in either direction, by the weather conditions and 'safety first' was the rule of the day, remembering *Titanic*'s ill-fated maiden voyage. Outwardly, Cunard officials repeated that the company was not in the business of racing but, defying that claim, on her sixth voyage, in August 1936, *Queen Mary* clocked up an average speed of 30.14 knots westbound and 30.63 knots eastbound, relegating *Normandie* to second place.

As explained in the previous chapter, the contest did not end there, though, for *Normandie* rose to the challenge. With new propellers, she bettered *Queen Mary*'s speeds in March and August 1937, raising the record to an average of 30.99 knots eastbound and 30.58 knots westbound. But the French repost had not reached its peak. In August 1938, *Normandie* improved on her record times even further by raising her eastbound average speed to 31.20 knots.

Queen Mary was, though, to have the last word in this unparalleled period of ocean rivalry. That same month, August 1938, she retook the westbound transatlantic record at 30.99 knots and on the return leg the eastbound record at 31.69 knots. The margins were small but *Queen Mary* was irrefutably the paramount express super-liner in a decade that had seen five different record-breakers and in which the transatlantic crossing time had been reduced by almost a day to fewer than four days. *Queen Mary* retained her crown through to July 1952, when she was finally usurped by *United States*. Her fourteen-year tenure as champion was only exceeded by the twenty years of the Cunarder *Mauretania* and the later seventeen active years of service of *United States*.

As the pain of the Great Depression gradually eased, Cunard were ready to proceed with *Queen Mary*'s consort and, on 6 October 1936, the contract for *Queen Elizabeth* was placed with John Brown & Co., Clydebank, as Yard No. 552. Though they were conceived as a matching pair, it is more correct to refer to *Queen Elizabeth* as the consort of *Queen Mary* rather than as her sister-ship. Advances in marine engineering and naval architecture in the five or so years since *Queen Mary* had been laid down permitted a number of conspicuous differences between the two *Queens*, the overall outcome giving them a dissimilar physical appearance.

Externally, *Queen Elizabeth* would have a more raked bow in order to stow a bower anchor, resulting in a longer hull length at 1,031ft. There would be no forward well deck and there was an absence of the four large box section ventilators on either side of *Queen Mary*'s upper deck. A reduction of boilers to twelve, half the number of *Queen Mary*, obviated the need for a third funnel. Her remaining two funnels also revealed a distinct variance in that they were to be supported internally, eliminating the requirement for external wire stays. All in all, the impression was of cleaner lines, perhaps still not as innovative as *Normandie*, but clearly *Queen Elizabeth* was to be a graceful, modern-looking thoroughbred that oozed power and stature.

Not only was *Queen Elizabeth* longer than *Queen Mary* but her tonnage was also greater, at 83,673 gross. Her beam, depth and draught measurements were more or less the same. No doubt her passenger numbers as initially planned would have been comparable but unfolding events left this open to future change.

Work on the second great Cunarder commenced with the laying of her keel on 4 December 1936. Construction proceeded

Queen Elizabeth on the slipway at John Brown's on 27 September 1938, about to be launched. (Cunard Archives, University of Liverpool)

Already repainted grey, *Queen Elizabeth* seen fitting out. (Glasgow University Archives)

smoothly and incident free, and she was ready to be launched in a ceremony performed by Her Majesty Queen Elizabeth, consort of King George VI, on 27 September 1938. All was progressing nicely to fulfil Cunard's grand plan for the two-ship, weekly express passenger service but once again the clouds of war were gathering over Europe. The entry into transatlantic service of *Queen Elizabeth*, intended for April 1940 to mark the Cunard centenary, had to be postponed – for six years as it turned out.

As a testimony to *Queen Mary*'s excellent performance on the North Atlantic run in those few remaining years of peace, by the summer of 1939 she had already completed some seventy round trips, averaging twenty-two per year and exceeding *Normandie* by more than 15 per cent. *Queen Mary* was profitable, whereas *Normandie* ran at a loss.

Queen Mary was in New York on the declaration of war with Germany and it was decided she should remain laid up there for her safety since the U-boat campaign had immediately commenced, as revealed by the sinking of the Anchor-Donaldson liner *Athenia* on 3 September 1939, just two days after the outbreak of hostilities. Only four days earlier, *Queen Mary* had left Southampton on her last pre-war departure, the crossing beset with anxiety and consternation among her passengers.

The two *Queens* constituted massively important military assets in support of Britain's war effort, able to convey vast numbers of troops to the theatres of war in the Middle East and Europe. Designated for the role of troopship on 1 March 1940, twenty days later *Queen Mary*, having been painted grey while moored in New York, sailed for Sydney, Australia, where she was to be fitted out to carry 5,500 servicemen, a fraction of what she would later transport. With her running mate, she was considered for conversion into an aircraft carrier under plans drawn up by the Admiralty in 1942, but that was as far as it went because their troop-carrying role was regarded as too valuable to lose.

Queen Elizabeth leaving the shipyard and heading to sea, a very different departure from Clydebank to that of *Queen Mary* four years earlier. (Glasgow University Archives)

Meanwhile, *Queen Elizabeth* was virtually ready to leave the shipyard, also to be taken up for trooping duties, but her presence on the Clyde was conspicuous and moving her to either Liverpool or Southampton could have placed her in equal danger. It was determined that it would be better to move her across the Atlantic to join her consort but it would have to be done under a cloak of secrecy in order to deceive any waiting German predators. Some work on fitting her out remained to be completed, so a number of workmen were instructed to sail with her on the pretext she was bound for Southampton. Each of them was issued with return rail tickets to Glasgow and they were all booked for the duration into hotels in the town, while parts and material for the outstanding work were shipped south by train. The subterfuge included fake preparations for her impending arrival. The King George V Graving Dock was made ready and tug crews were placed on standby. Perhaps the most audacious part of the ruse involved taking aboard the Southampton Harbour pilot at Clydebank, no doubt to his utter astonishment when he subsequently discovered where he was actually headed.

All the duplicitous arrangements were made covertly but with enough information slipped out in 'idle gossip' to fool anyone involved in espionage for Germany. A major hazard was the dependence on high tide to provide sufficient bottom clearance as *Queen Elizabeth* made her way down river to the open sea, a factor that could not have escaped the attention of the Germans. Nonetheless, she was already painted naval grey as she set off on 26 February 1940, and the manoeuvre was undertaken without a hitch. On entering the Firth of Clyde, her master, Captain John Townley, opened sealed orders revealing her sailing date of 3 March and her true destination.

Six days later the 'Grey Ghost', as she was dubbed, arrived at New York, where she joined *Queen Mary* and *Normandie*, the only time these three great ships were to be in port together. Coincidentally – or perhaps not – Luftwaffe bombers attacked the docks at Southampton on the day she had been due to arrive there.

Queen Elizabeth heading across the Atlantic on her first voyage. It was 'silent running' amid the secrecy surrounding her wartime crossing. (World Ship Photo Library)

Queen Elizabeth in Sydney Harbour on 21 February 1941 showing her with her hull repainted black. The dividing line with the grey-painted upperworks is a deck higher than it will be later when her hull is repainted black in her civilian livery. (SuperStock / Alamy Stock Photo)

RMS *Queen Mary* as a Second World War troopship in December 1944, showing her degaussing coil. This was how she would have been seen by crew aboard HMS *Curacoa* as the liner bore down upon the cruiser on 2 October 1942. It is said that Winston Churchill signed the D-Day declaration aboard her in the suite he called his 'War HQ at sea'. (Warfare History Network)

Queen Elizabeth had not undergone any of the trials that were usually conducted pre-handover and she was in all respects totally untried and untested. Despite this, she had performed superbly on the crossing and presented no concerns whatever to her hastily signed-on scratch crew. During the voyage she achieved the astounding maximum speed of 36.25 knots!

The wartime record of the *Queens* was exceptional, a facet of their career stories for which they are renowned. Both ships packed a huge amount of activity into the six years of war, such that it is relevant here to provide only a concise account of the main highlights of their incredible troop-carrying exploits.

Queen Elizabeth headed for Sydney for her conversion on 13 November 1940, calling en route at Port of Spain, Trinidad, and Simonstown, South Africa, and Singapore. At Singapore, for some unexplained reason, her hull was repainted black, giving her a two-tone colour scheme with her grey upperworks. This was not retained for long, although it is not known precisely when she reverted to grey all over.

Six months earlier, *Queen Mary* left Sydney on 5 May 1940 on her initial trooping voyage, bound for Gourock in convoy with two Cunard fleet mates, *Aquitania* and *Mauretania* (second of the name), along with three Canadian Pacific liners, the *Empresses of Britain*, *Canada* and *Japan*. The first trooping voyage of *Queen Elizabeth* commenced from Sydney on 11 April 1941, taking her in convoy via Fremantle and Trincomalee to Suez, returning via Singapore.

Thereafter, together the *Queens* maintained a shuttle between Sydney and Port Tewfik, Suez, through to mid-1942, conveying Anzac troops, after which they were switched to the North Atlantic delivering GIs to Britain by the notorious shuttle, in readiness for Operation Overlord, the invasion of Normandy. They sailed in convoys but were considered to be safest when they operated independently, able to rely on their high speed to outrun any attackers. Both ships had been made the subject of bounties as an inducement to enemy submariners to make particular effort to sink them. Moreover, the Suez run was fraught with hazards, especially in the final approaches through the Red Sea where they were exposed to Luftwaffe attacks as had occurred to *Georgic*, another Cunard White Star ship, on 7 July 1941.

Between them, they clocked up over a million miles as troopships, 664,771 by *Queen Mary* and 492,635 by *Queen Elizabeth*. Within this mileage, there were some epic voyages. On 18 February 1942, *Queen Mary* sailed from Boston with 8,938 American troops on what came to be called the '40 Days and 40 Nights' voyage. She arrived at Sydney on 28 March. Later that year, on 23 December 1942, she set sail from Gourock to Freetown, Cape Town, Aden, Suez and Sydney, returning to Gourock on 22 April 1943. Dubbed 'the Long Voyage', it covered a total of 37,943 miles.

The immense distances travelled in each case are reflected in the equally vast number of military passengers conveyed by the pair. By the end of their respective auxiliary careers, *Queen Mary* had carried 810,730 enlisted personnel and *Queen Elizabeth*'s total, slightly higher, was 811,324. Such levels of military manpower had a profound effect on the course of the war and, by a prodigious margin, the *Queens*' accomplishment was unrivalled by any other vessel engaged for transportation duties in the conflict.

Queen Mary was distinguished for taking the Prime Minister, Winston Churchill (under the name Colonel Warden), across the Atlantic on three occasions in connection with Allied conferences. Between 5 and 11 May 1943 he crossed aboard her from Gourock to New York along with some 5,000 German prisoners of war. The next occurrence was a voyage to Halifax from Gourock, from 5 to 9 August 1943. After the conference there, on the return crossing to Gourock, the Prime Minister was able to experience at first hand the extraordinary troop-carrying capability of Cunard's *Queens*. Sailing on 27 August and arriving five days later, on board with him were 15,116 US troops.

This immense number had, in fact, been surpassed on an eastbound crossing completed on 30 July 1943 in which *Queen Mary* carried a record total complement of 16,683 persons, a total that has never been exceeded since. It comprised 15,740 troops and other military personnel and 943 crew, a relatively small number in contrast, who would have been kept very busy! *Queen Elizabet*h's greatest wartime complement added up to a combined total of 15,932 troops and crew members.

During the course of their auxiliary service both ships were fitted with anti-aircraft guns, installed forward, later supplemented with heavier-calibre weapons, none of which were fired in anger. They also received mine countermeasure equipment, first paravanes, between August and September 1940 for *Queen Mary* and between May and June 1941 for *Queen Elizabeth*, the work carried out in drydock at Singapore. Subsequently, degaussing strips were installed at the Cockatoo Island Shipyard, Sydney, fitted to *Queen Mary* in December 1940 and in early 1941 to *Queen Elizabeth*. The coils remained as discernible features on their hulls for the remainder of the war. The Japanese invasion of Singapore in February 1942 denied use of the drydock there to the British and increased reliance on the Cockatoo Docks and Engineering facility. This was the only dockyard with drydock and repair capabilities in the South-West Pacific and was essential to the operation of the *Queens* until North American drydocks became available after the United States' entry into the war in December 1941. It should be noted that because of these difficulties, *Queen Elizabeth* still had parts of the launch cradle attached to her underwater hull a year after she had crossed to New York and two years from when she had been launched.

In April 1941, the *Queens* sailed together as troopships for the first time in Convoy US 10, taking them from Sydney to Fremantle, Trincomalee and Suez, returning via Singapore. A year later, on 4 June 1942, *Queen Elizabeth* left New York bound for the Clyde in Convoy AT 15Z initiating the 'GI Shuttle' that, apart from a run to Suez, Cape Town and Rio de Janeiro between 17 June and 19 August as part of Convoy WS 19Y, she remained on until the end of the war. *Queen Mary* joined her from May 1943. As a measure of the *Queens* contribution to the military build-up for the D-Day invasion and in sustaining reinforcements for the Allied forces, as the campaign to defeat Germany drove onwards, *Queen Elizabeth* alone crossed the Atlantic ninety-eight times in forty-five months, with only a single dockyard visit at Boston in November 1944.

The *Queens*' phenomenal wartime careers were only blighted by an unfortunate collision between *Queen Mary* and her light cruiser escort HMS *Curacoa* on 2 October 1942. The great liner, carrying in excess of 11,000 American troops, rammed and cut in half the far smaller naval vessel. The tragic incident remained confidential throughout the continuing hostilities for obvious reasons until, after the war, it assumed the prominence its gravity warranted. In January 1947, the Admiralty brought a legal action against the Cunard White Star Line for damages, claiming negligence as the cause of the collision.

The troopship *Queen Elizabeth* looking resplendent in navy grey and revealing her degaussing coil. Note the guns mounted aft. (Authors' collection)

Arriving at New York on 16 July 1945, *Queen Mary* crowded with 14,526 GIs returning from Europe. Many would be redeployed to the Pacific. (Isaac Bethel Utley, Office of War Information)

In essence, the collision arose when the ships' close manoeuvres brought them onto a converging course and *Curacoa* crossed the bow of the onrushing troopship then steaming at 26.5 knots. At the time *Curacoa* was making 25 knots. She had met *Queen Mary* at 0900 hours that morning 40 miles north of Tory Island, off the coast of Northern Ireland, her duty being to escort the troopship and provide anti-aircraft protection for the remainder of her voyage to the Clyde Northwest Approaches. The troopship was following zigzag pattern No. 8 as previously agreed as being the most suitable between the two commanding officers, Commodore Sir Cyril Gordon Illingworth of *Queen Mary* and Captain John Wilfred Boutwood of HMS *Curacoa*, and in accordance with orders issued in New York by the Operational Authority responsible for transatlantic trooping voyages.

There was a heavy following swell that tended to cause the two ships to yaw, leading to some confusion as to whether they were holding a steady course or changing their bearings. A further complication was that the U-boat U-407 had been detected in the area, posing a threat to *Queen Mary* and requiring the cruiser to attempt to prevent it from making a closer approach or taking up an attack position.

Initially 10 miles apart, bearing changes brought the vessels to a point where, at 1400 hours, they were only a mile apart with *Curacoa* off *Queen Mary*'s starboard side, steering a parallel course. At 1404 *Queen Mary* made the next course alteration in her zigzag pattern. Three minutes later the ships were converging, fast approaching each other at around a fifth of a nautical mile (405 yards) each minute. *Curacoa* took evading action but it was too late and at 1412 *Queen Mary* struck her on the port side, 300ft from her bow, cutting her in two. *Curacoa* sank almost immediately, 338 members of her crew losing their lives.

Queen Mary was unable to stop to assist, indeed she was not permitted to take such action because it would have exposed her troops to danger. Several destroyers were in the vicinity, the Royal Navy ships HMS *Bulldog*, *Bramham*, *Cowdray* and *Skate* and the Polish ORP *Blyskawica*, and these went to the aid of survivors, rescuing 101 sailors from the sea.

Queen Mary suffered a twisted and fractured stem but no other damage. A temporary repair was performed at Gourock, plugging the aperture with 70 tons of cement. After returning across the Atlantic, a more permanent repair was hastily carried out at the Boston Navy Yard in the form of a more durable bow section. Post-war, *Queen Mary*'s bow plating was fully restored.

The Admiralty's legal case five years later, which asserted dereliction of duty on the part of *Queen Mary*'s officers of the watch, was heard in the Admiralty Probate and Divorce Court. However, Justice Sir Gonne St. Clair Pilcher, the presiding barrister, found *Curacoa* to be wholly to blame for the collision, fully exonerating the Cunarder. The decision was not received well by the Admiralty and they took the case to the Court of Appeal, where the fault was apportioned two thirds against *Curacoa*, one third against *Queen Mary*. Yet again, unhappy with this attribution of blame, the Admiralty appealed a second time, taking their case to a House of Lords tribunal on 8 February 1949. Nonetheless, the five Law Lords of the Appellate Committee decided that the Court of Appeal's judgement should be upheld and that was the end of the matter, not that the outcome was of much consolation to the many bereaved families of the disaster's naval victims.

As part of the ruling's justification, it was accepted in the hearings that it was incumbent upon the naval escort to keep clear of the troopship because of her limited ability to take effective evasive action at such short distances. Furthermore, it was concluded that watch-keeping aboard the cruiser had been inadequate, besides which she had the primary responsibility to give way, even though there was uncertainty as to the applicability of the 'Rules of the Road' during wartime.

After the war was over in Europe, *Queen Mary* made her first return to her home port, Southampton, on 11 August 1945, after six years of absence. *Queen Elizabeth*'s long-awaited first entry into the port occurred nine days later. *Queen Elizabeth* was the first of the pair to be paid off from government service. After a full overhaul of her engines on the Clyde and a comprehensive refit of her interiors at Southampton,

Post-war, *Queen Elizabeth* is finally put through her commercial sea trials in the Firth of Clyde. (Glasgow University Archives)

she underwent official trials six years later than had been intended. If anything, she was the faster of the two and was quite capable of taking the Atlantic Blue Riband record from *Queen Mary*. But Sir Percy Bates would have none of it. Cunard would not, he insisted, sanction any speed competition between two of its own ships, a directive that resulted in *Queen Elizabeth* being denied the opportunity to show what she was truly capable of on the North Atlantic run.

Having been granted the distinction of making Cunard's first post-war transatlantic voyage, thereby instating the company's two-ship express service, *Queen Elizabeth* made her maiden commercial sailing from Southampton on 16 October 1946. In the meantime, while *Queen Elizabeth* was being prepared, *Queen Mary* continued about her auxiliary duties until September 1946, first repatriating US and Canadian servicemen, followed by the conveyance of a total of 12,886 war brides to North America with their children. Through this period, *Queen Mary* retained her navy grey livery with the exception of her funnels, which had been repainted in Cunard red and black on 29 August 1945.

Like *Queen Elizabeth*, *Queen Mary* underwent a thorough, year-long overhaul and refit that resulted, in her case, in an increase of her gross tonnage to 81,237. Following the interval of the war, the passenger grades on North Atlantic ships had been revised, reverting partially to how they had been before 1936 when, on the behest of Cunard, the cabin-class category had been introduced. The *Queens'* post-war passenger breakdowns reflected this. Thus, *Queen Mary*'s passenger numbers, which had also been modified, became 711 first class, 707 cabin class and 577 tourist class. As for *Queen Elizabeth*, her figures were 823 first, 662 cabin and 798 tourist.

Queen Mary made her inaugural post-war commercial departure from Southampton on 31 July 1947. At long last, after twenty years' gestation from when it was first conceived, Cunard's fully fledged two-ship weekly service was up and running. By then, though, there were emerging issues that would confront the *Queens* to ultimately undermine their combined operation.

Within five years, there was a new challenger on the scene when the American-flag *United States* entered service on 3 July 1952, immediately taking the Atlantic Blue Riband from *Queen Mary* in both directions and by the greatest margin ever in the history of Atlantic speed rivalry.

As the decade of the 1950s advanced, the demand for international civil aviation increased relentlessly, employing aircraft whose design had benefitted from accelerating technical advances stimulated by wartime developments. Foremost was the invention of jet propulsion, which, when integrated into commercial passenger airline design, resulted in a step increase in air speeds, spelling an impending challenge to travel by sea.

In the early years of her commercial career, *Queen Elizabeth* arrives at Southampton accompanied by Red Funnel and Alexandra Towing tugs. (Adrian Vicary)

Another view of *Queen Mary* in The Solent taken from her port bow in this aerial view. Note that in this post-war photograph she carries radar. (Authors' collection)

Seen in The Solent, *Queen Mary* from her starboard stern quarter. (Authors' collection)

The *Queens* held their own until the early 1960s but the ominous signs were clear to see as passenger revenues fell progressively season after season. By this time, the life expectancy of the *Queens* was also a matter for deliberation in the Cunard Line boardroom. The French Line had already committed to the construction of a new three-class flagship, *France*, which it was intended should follow the traditional all-year-round transatlantic service formula. Possibly in light of this, Cunard also opted, initially, for the same long-established operational approach, reckoning perhaps on there being sufficient traffic volumes to support both ship and aircraft services. Thus, a design for a modern version of the *Queens* was pursued, dubbed *Q3*, also of three-class layout but of slightly lower tonnage at 75,000 gross and slightly shorter length at 990ft overall. However, in light of continuing developments, this concept fell by the wayside, but that is going ahead of the *Queens*' story and it is detailed fully in the chapter on *Queen Elizabeth 2*.

Back in the 1950s, a major concern had been passenger comfort. When *Queen Mary* entered service she had some air conditioning, limited to the public spaces, and *Queen Elizabeth* was equipped likewise when she first entered service. Full air conditioning, including to the passenger cabins, came later for *Queen Elizabeth*. Of greater consequence was their tendency to roll in a seaway, sometimes excessively, particularly in North Atlantic winter conditions, for which they had something of a reputation. Indeed, *Queen Mary* was so inclined to roll that one crew member reputedly said 'she was so tender, she could roll the milk out of a cup of tea'. It was an uncomfortable experience at best but in heavy seas one that could be disconcerting to the unseasoned traveller. This impediment was rectified when they were fitted with retractable Denny Brown fin stabilisers installed by Vosper-Thornycroft in the King George V drydock at Southampton. *Queen Elizabeth* received hers, two on each side, in January 1955 during her winter overhaul, and *Queen Mary* in March two years later, the installation in her case being more complicated due to the amount of engine room space occupied by her greater number of boilers. This required the control mechanisms to be mounted vertically rather than horizontally.

The unfolding years from 1946 to the early 1960s were steady ones for the two *Queens*, largely uneventful apart from groundings on at least two occasions, on the Brambles Bank in The Solent and off Cherbourg. A potentially more serious incident involved *Queen Elizabeth* when she collided with the US cargo ship *American Hunter* off Coney Island, in the entrance to New York harbour, on 29 July 1959. Uncharacteristically for the time of year, there was dense fog with virtually zero visibility. It was uncomfortably reminiscent of the tragic collision three years earlier between the liners *Andrea Doria* and *Stockholm*.

Repeatedly sounding her foghorn, *Queen Elizabeth*, under command of Commodore George Morris and carrying 1,959 passengers, was outbound for Southampton. She had cleared The Narrows to enter the Lower Bay, heading south-east to pass the Ambrose light vessel when she was struck on her starboard bow by the cargo ship. *American Hunter*, under Captain George Walker, was returning from France and the UK; she sustained a crumpled bow and a 20ft dent in her side. Fortunately, the damage to *Queen Elizabeth* was only minor and there were no casualties. Both ships carried radar and had been proceeding under the direction of pilots who, it was later reported, claimed the collision was unavoidable.

Tugs, coastguard vessels and other rescue craft were hastily despatched to the scene because the vessels had hove to in the middle of the busy channel and were completely concealed from view. As a precaution, the Port of New York Authority closed the Ambrose Channel to all ships until the damaged vessels had been removed or had manoeuvred away under their own steam. Unassisted, *Queen Elizabeth* returned to her New York pier, where repairs were carried out expeditiously while she was alongside, the outcome ultimately being nothing more than a delay of several hours to her departure.

By the early 1960s, the competition across the Atlantic from aircraft was biting ever deeper into the *Queens*' revenues and Cunard's profits. With a replacement vessel still some years off and Cunard's determination to keep the older *Queens* operational into the latter part of the decade, they were switched increasingly

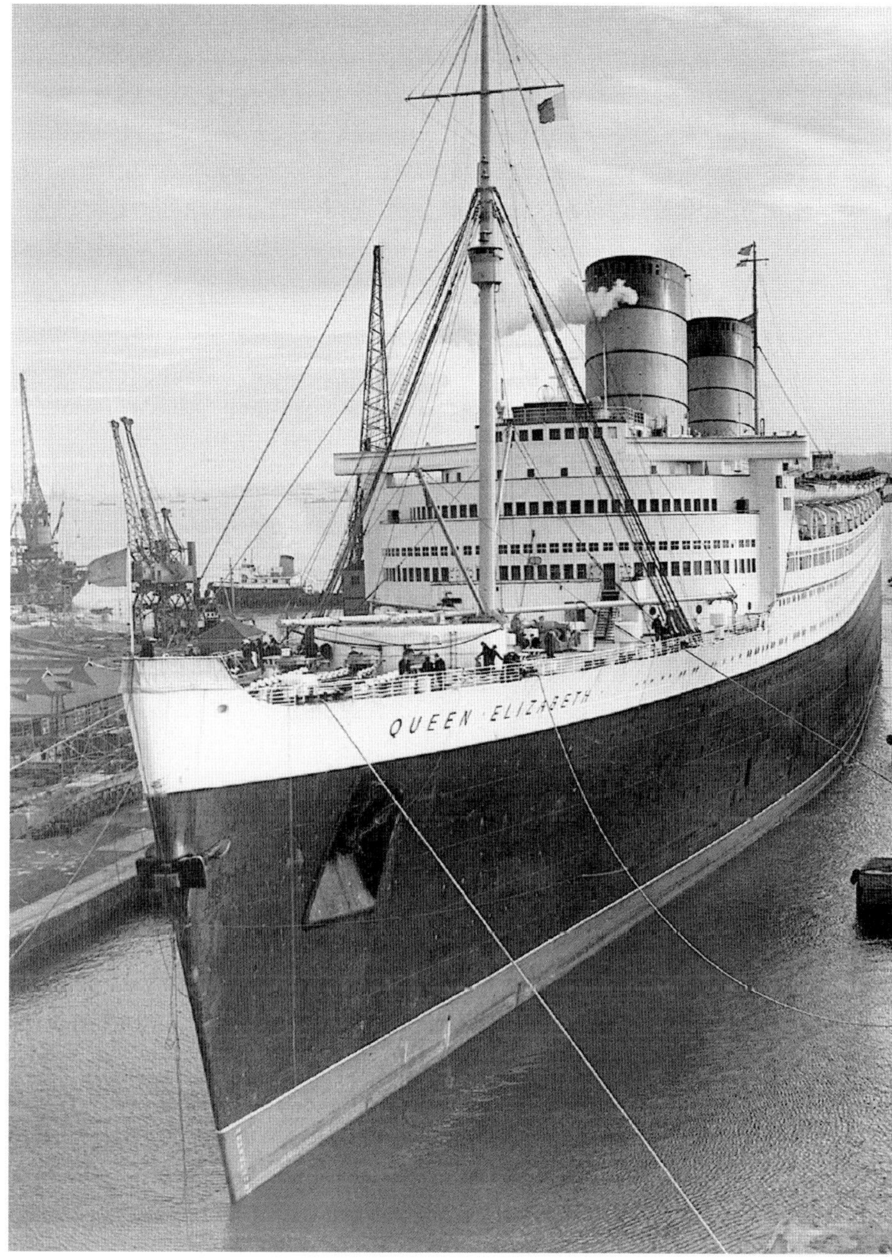

Queen Elizabeth assisted from the King George V drydock after one of her regular overhauls. (British Railways, Docks & Marine)

to off-season cruising, short trips across the English Channel and longer excursions to the Mediterranean or from New York to Nassau in the Bahamas. They were totally unsuited for this work, however, lacking the outdoor recreational amenities expected by cruise passengers.

To at least make *Queen Elizabeth* more appealing in the cruise market, she was taken in hand in 1965 to undergo a comprehensive, life-extending refit. The objective was to retain her in service until 1975 to operate eventually in conjunction with the new *Queen Elizabeth 2*, by then under construction at Clydebank, when she entered service in 1969. Given her age, then twenty-five years, and the uncertainty of her future prospects, the investment in *Queen Elizabeth*'s adaptation was ambitious, the work involving the addition of a lido deck aft with an outdoor swimming pool and the extension of air conditioning throughout her cabin accommodation. On completion, she had a revised registered tonnage of 82,997 gross.

Arriving at Southampton, *Queen Elizabeth* assisted by the tugs *Clausentum* and *Romsey* and the Red Funnel tender *Gatcombe*. (Tom Rayner)

However, any hope that this strategy would halt the *Queens'* declining fortunes was forlorn and dashed within twelve months. The British seamen's strike, from May to June 1966, proved to be the final nail in the coffin, occurring just weeks after the completion of *Queen Elizabeth*'s refit.

On 8 May 1967, Cunard announced that *Queen Mary* would be withdrawn from service that October after thirty-one arduous years on the North Atlantic run and as a troopship. She departed New York on her final transatlantic voyage on 22 September 1967. Sold to the Corporation of Long Beach, California, for £1.23 million, to become a static museum, hotel and convention centre, she was to be preserved rather than broken up. But this was only a partial consolation to her many admirers who thronged Southampton and the shorelines along The Solent to see her off on 31 October 1967. With Commodore John Treasure Jones in command, *Queen Mary*'s final voyage was a long, fare-paying cruise down the length of the Atlantic, around Cape Horn and returning north along the Pacific coast of South and Central America. She called at Lisbon, Las Palmas, Rio de Janeiro, Valparaiso, Callao, Balboa and Acapulco before she arrived at Long Beach on 9 December 1967, ending one of the longest voyages she had ever undertaken, powered throughout by her own engines.

An in-depth conversion followed in which a number of her original fittings, including her corroded funnel casings, were removed and replaced by synthetic equivalents. It took three and a half years to complete. On 27 February 1971, she was moved from Pier E at Long Beach to Pier J, where she has remained ever since, opening to the public for the first time on 10 May that year. For ten years from 1983, she formed part of a combined tourist attraction with Howard Hughes' giant flying boat, the H-4 Hercules, known as the 'Spruce Goose'. Despite her enduring popularity, *Queen Mary* has represented an expensive commitment to her new owners from a maintenance angle, as well as through fluctuating visitor numbers, and over the more than forty years of her Californian sojourn she has experienced repeated periods of uncertainty, more than once threatened by closure and disposal. However, by the third decade of the twenty-first century, at almost 90 years of age, she still remained functional and it is to be hoped she will remain so for many more years.

The fading aspirations that *Queen Elizabeth* would be spared the axe for at least eight more years were short-lived and in the year following *Queen Mary*'s sale she made her final Cunard voyage from New York on 30 October 1968, ending her days as a super-liner. Two days short of a month later, she sailed to Port Everglades, Florida, where she was to be adapted to perform a similar role to that of *Queen Mary* at Long Beach. She had been sold for £3.25 million to a consortium of Philadelphia businessmen, the Elizabeth Corporation, although Cunard retained a controlling 85 per cent interest in the enterprise. Upon her arrival on 8 December 1968, she was initially moored at a temporary berth, the intention being to move her to a dock in the Broward County deepwater port. Allowing her to be partially flooded, to sit on the bottom at Fort Lauderdale, was also considered. Although a substantial refurbishment was planned, she was opened to the public in February 1969 after the shortest possible preparation for her static role. Instigating a precarious period for the grand old liner, she was sold to Queen Limited on 19 July 1969 and her name shortened to *Elizabeth* but within the year that operation had failed, losing money, and she was closed down prematurely when the port authorities declared her to be a fire hazard.

A year later, in August 1970, she was sold by auction and acquired by the Hong Kong shipping magnate Tung Chao-yung (known as C.Y. Tung), who owned the Orient Overseas Line. She was to be converted into a specialist sea-going university cruise ship, a 'world campus afloat' as it was called, soon rejigged as 'semester at sea'. At the time it was heralded as the 'rebirth of a legend!' But she had deteriorated in the months since her transfer to Port Everglades and the passage to Hong Kong, where the conversion was to take place, proved to be a protracted affair beset with engine failures, fires and other machinery problems. Commodore Geoffrey Thrippleton Marr, who was engaged to take command for the voyage, had recommended that a tow to the Far East would have been preferable. Departing from Port Everglades on 10 February 1971, calling by necessity en route at Rio de Janeiro, Cape Town and Singapore, she reached Hong Kong on 16 July 1971, where she anchored of Tsing Yi island before being moved nearer to the entrance of Victoria Harbour, south of Stonecutters Island.

She had been renamed *Seawise University* before leaving the United States, a name that in itself signified a distinct change to her character, far removed from her transatlantic days, and so it would have been. A £5 million adaptation ensued to convert her into a luxurious, all first class cruise ship for 1,400 passengers. On completion, she was to be moved to Japan for drydocking and then to Los Angeles for the commencement of her inaugural 'Circle Pacific Cruise'. Intended to commence on 28 March 1972, it would take in Honolulu, Suva, Sydney, Fremantle, Bali, Singapore, Hong Kong, Kobe, Yokohama, Honolulu (second call) and Vancouver, concluding with her return to Los Angeles. It seemed that the sea-going campus venture had been put on hold for eventual introduction at some later, unspecified date. In the event, neither the cruise ship nor the floating university role came to pass.

By January 1972, the conversion, which had started on 16 July 1971, was nearing completion and the former *Queen Elizabeth* was looking resplendent in her new all-white livery with yellow funnels, upon which were the distinctive, stylised red and white flower symbols of the Orient Overseas Line. Sea trials under her new commander, Commodore Chen Ching-yien, were scheduled for the middle of the month. On the afternoon of 9 January, to celebrate her impending return to active service, a cocktail party was to be held aboard to entertain a large group of guests, some sixty of Hong Kong's most elite business people. Unknown to them, a combination of lax security, the lack of a tested fire prevention plan and inadequate firefighting personnel contributed to threaten the festivities.

Responsible for the conversion was Island Navigation Corporation, an affiliated C.Y. Tung Group company, from its repair and maintenance yard at Lai Chi Kok. On board on that fateful day were thirty-one of the company's workmen along with 263 sub-contractors. There were also around 205 members of *Seawise University*'s crew, making a total with the guests who had already arrived of approximately 500. Among the crew and workmen there were only four who had been allocated fire patrol duties, plus four professionally trained firemen for the entire ship. An additional nineteen fire-watchers were posted at vulnerable positions where numerous welding operations were in progress at various sites on Decks A to D, plus on Decks M, P and R.

Fire was first discovered on A Deck at 10.30 a.m. on 9 January by three junior crew members, who raised the alarm. Fanned by strong winds crossing the harbour, the fire increased in intensity and rapidly began to spread, but action to extinguish it was frustrated by a convoluted chain of command – for example, a fire officer order to a deck cadet to find and report to the duty officer, who in turn had to communicate with a chief officer, who finally transmitted an emergency call. Thereafter, the public address system failed, inhibiting further emergency announcements. Meanwhile, another chief officer prematurely and unilaterally ordered the lowering of lifeboats, while simultaneously the commodore was informed of more fires around the ship, on the sun deck and on B deck. Ultimately, at least four or five other fires were located, with outbreaks on R deck, D deck as well as others, previously unseen, on B deck. So many random fires clearly indicated an act of sabotage.

In fact, *Seawise University* was well-equipped for fire prevention and fighting. She had smoke detector alarms, a comprehensive sprinkler system and her appliances included numerous hydrants, hoses, portable extinguishers and sand buckets, as well as breathing apparatus, axes, crowbars and other tools. Despite this, the firefighting was not well co-ordinated and the fact that numerous portholes and shell doors were left open served to help the fires to spread. Added to that, dealing with so many eruptions in so many places in these circumstances would, of course, have seriously tested the most efficient and effective response.

There was, though, a hint of the disorder and turmoil that had characterised the fire that destroyed *Normandie* fifty years earlier. Just after midday, an appeal was made by the Hong Kong Fire Service Department for all harbour craft to render assistance to tackle the fire that by then was raging throughout the ship. It was not long before the blazing *Seawise University* was surrounded by a variety of harbour and police launches and a large number of fireboats, all directing water onto the portside superstructure under the direction of a command post set up on the fireboat *Alexander Grantham*. This continued throughout the next fifteen hours, during which time, at 12.55 p.m., firefighters boarded the stricken vessel to search for any survivors who had not made their escape and to help tackle fires close to wherever possible. Gradually the ship began to list to starboard, a process that accelerated as water poured into her through the open shell doors on that side of her hull.

By the following day, *Seawise University* was in a sorry state, clearly beyond recovery. Her upper decks had collapsed and finally, as her list increased, she rolled over onto her starboard side, her portside partially above the surface. It was a devastating end for one of the greatest liners ever built. Fortunately, there had been no human casualties.

A Marine Court investigation into the disaster took place in Hong Kong's Legislative Council Chamber from 8 to 30 February under the jurisdiction of Puisne Judge Art Michael McMullin. Its report was published on 29 June 1972, finding many reasons to criticise the owners and the refurbishment contractors but concluding that the cause of the fires was arson perpetrated by a person or persons unknown. There was speculation and unsubstantiated rumours as to the motivation – fraud, union disputes, even dissension between Nationalist and Communist elements of the workforce. But, in the aftermath, all that was completely academic. The Royal Hong Kong Police pursued enquiries but no charges were ever made. The ship was irretrievably lost and the wreckage was cleared to remove a potential hazard to other shipping, the process taking around twelve months and all but complete by the end of 1975.

Tugs stand by to assist as *Queen Mary* approaches Southampton Docks. (Sueddeutsche Zeitung Photo / Alamy Stock Photo)

If there was ever an example of what was intended by the commissioning of the great North Atlantic super-liners, *Queen Mary* and *Queen Elizabeth* had indubitably been it. Despite the interruption of the Second World War, together they achieved greater passenger figures than all their predecessors combined. Over her career, *Queen Mary* had travelled 3.6 million miles carrying a total of 2,285,000 passengers. As for *Queen Elizabeth*, she had steamed 3.2 million career miles with 2,300,000 passengers. These figures were only exceeded subsequently by *Queen Elizabeth 2* but, in her case, it was over a longer period.

It was once said of Cunard's *Queens*, 'They were each superbly well-crafted for a particular, and very specific, task – ferrying people in great style forwards and backwards across the North Atlantic. But when all that came to an end, there wasn't really very much either of them was suited to do.'

QUEEN MARY

Opposite: Queen Mary entering King George V drydock on 16 July 1948. (British Railways, Docks & Marine)

Below: Underway in The Solent. *Queen Mary* remained the most popular of the two *Queens* throughout her career. (Richard de Kerbrech collection)

Queen Mary making slow progress along Southampton Water bound for the docks. (Fotoflite/ Nigel Scutt)

Already renamed *Seawise University*, the former *Queen Elizabeth* called at Cape Town en route to Hong Kong. (Adrian Vicary)

Seen after arrival at Long Beach, California. Since then, *Queen Mary* has spent more than twice the time as a hotel, museum and convention centre than she did as a serving Cunard liner. (Long Beach Corporation)

The former *Queen Elizabeth* under conversion as *Seawise University*, her funnels temporarily painted with a red primer prior to receiving the white Orient Overseas colour scheme. (Ian Shiffman)

Above: Seawise University engulfed with fire on 9 January 1972, its cause a senseless act of sabotage. Within the smoke, the flower emblem of the Orient Overseas Group can still be seen on the forward funnel. (Authors' collection)

Left: The wreck of *Seawise University*, yet another super-liner lost to the ravages of a raging inferno. (Bill Miller collection)

Above: A Hamburg Amerika Line poster advertising cruises pre-First World War, showing an imposing image of *Imperator*. The poster advertises her entry into the transatlantic service as 28 May when she actually commenced her maiden voyage on 10 June 1913. (Hapag-Lloyd)

Above: A poster advertising the United States Lines' *Leviathan*. She was in fact operated on behalf of the US Shipping Board. Although she had distinguished herself with yeomanry service during the First World War, her commercial service suffered during the Prohibition era. (United States Lines)

Left: A colour painting of *Berengaria*, Cunard's flagship. She operated in Cunard's 'express' service at a service speed of 24 knots. This compared favourably with *Mauretania*'s 26.5 knots and the *Aquitania*'s 23 knots. (Authors' Collection)

WHITE STAR LINE
"MAJESTIC"
THE LARGEST STEAMER IN THE WORLD
LENGTH – 954·5
BREADTH – 100·15
DEPTH – 64·2 FT.

Opposite page:

Above left: An aptly named super-liner, she was nicknamed the 'Magic Stick'. *Majestic* was the world's largest liner until 1935. Along with *Homeric* and *Olympic*, she maintained White Star's regular service and was its first turbine-driven liner. (White Star Line)

Above right: Norddeutscher Lloyd's *Bremen* in an image created to impress upon the travelling public that speed was the order of the day. (Deutsches Schifffahrtsmuseum)

Below: Europa became French Line's *Liberté*. It was not quite up to the standard of *Normandie* but it gained popularity among the travelling public and was also featured in the film *The French Line*. (Authors' collection)

This page, below: A striking painting of Italia's *Rex*, showing her inward bound for New York passing the Ambrose Light Vessel at the end of her record-breaking crossing. She is seen flying the Blue Riband pennant from her mainmast and sounding her siren to mark the achievement. (© Stephen J. Card AFNI)

SHIPPING WONDERS
OF THE WORLD
PART
33
7D
WEEKLY
REX

NORMANDIE

Opposite page:

Above left: From the cover of a pre-war shipping magazine, an impressive sea level view of *Rex* bow-on at speed. (Authors' collection)

Below left: *Normandie* alongside at Pier 88 in New York's 'Luxury Liner Row' at the end of her commercial career. Her massive size eclipsed that of *Majestic*. (Bill Miller collection)

Right: A painted impression of *Normandie* viewed from her bow. From this angle the magnitude of this ship of state can be fully appreciated as well as her graceful bow lines. (Authors' collection)

This page, right: *Queen Mary* being eased into her berth alongside Southampton's Ocean Terminal. Until her entry into service in 1936, Cunard had not ordered a liner of such significant size in more than twenty years – not since *Aquitania*. (Phil Fricker)

Queen Mary is seen making her final departure from Southampton on 31 October 1967, blowing her steam whistle and with Red Funnel's *Gatcombe* in attendance. (Kenneth Wightman)

Cunard's *Queen Elizabeth* outward bound from Southampton for New York. She was the largest passenger liner ever built in the UK. (Mike Lindsay)

Queen Elizabeth outward bound in Southampton Water. (Phil Fricker)

Right: The first post-war-built super-liner, United States Lines' *United States* seen here in The Solent on 21 August 1969, outward bound for New York, on her last call at Southampton. (Phil Fricker)

Below: United States alongside Southampton's original Ocean Terminal. (Don Smith)

Above: Whenever the berth at Southampton's iconic Ocean Terminal was occupied by *Queen Mary* or *Queen Elizabeth*, the *United States* was obliged to moor nearby at the port's Western Docks. She is seen here departing on a quiet day, possibly in April 1959. There seems to be a lack of activity and fanfare on the quay on this occasion. (British Transport Commission)

Opposite above: The world's fastest liner and the last 'true' holder of the Blue Riband 'stretching her legs' in the English Channel. *United States* seen in her element. (Nigel Scutt/Fotoflite)

Opposite below: Moored either side of their Hoboken quay in May 1919, the US troopships *Imperator* (left) and *Leviathan*, ex *Vaterland* (right), at the time the world's two largest ships. (US Naval Historical & Heritage Command NH103126)

THE "GIANTS" of THE SEA
NEW YORK, N.Y.

Luxury Liner Row, *Manhattan*, in the 1930s showing four of the super-liners in dock. From top left, the ships depicted are *Europa*, *Rex*, *Normandie*, *Georgic* of White Star Line and *Berengaria*. (Authors' collection)

Europa passes Southampton's Ocean Dock outbound during the 1930s. An interesting aspect of this picture is that all the ships seen, with the exception of the twin-funnelled *Arandora Star* (centre right), are of German origin, sequestrated by Britain after the First World War. At the Ocean terminal is *Majestic*, built as *Bismarck*, at 46 Berth (left) is *Homeric*, ex *Columbus*, and in the Trafalgar drydock (extreme left) is *Empress of Australia*, formerly the Hamburg Amerika Line's *Tirpitz*. (KGPA Ltd / Alamy Stock Photo)

A 1950s view of Southampton's Eastern Docks. *Queen Elizabeth* is berthed at the Ocean Terminal. *United States* is passing in the foreground, heading for the Western Docks. (British Railways, Docks & Marine)

Le Havre on 23 November 1961 on the occasion of *France*'s first arrival at her homeport, passing *Liberté* moored at the Quai Joannes Couvert, the home of the port's Gare Maritime. The occasion represented a symbolic as well as actual transfer of flagship status between the two great French super-liners. (Jean-Yves Brouard collection)

Left: An imposing photo of *France* taken on 6 September 1967 as she rounds the West Bramble Buoy in The Solent, returning to Le Havre. (Phil Fricker)

Below: France being eased into Southampton's Ocean Terminal. A study in early 1960s streamlining. (Chris Howell)

Above: From the world's largest liner as *France* to the world's largest cruise ship: *Norway* sporting her new livery at Hamburg on 18 August 1984. (Wolfgang Fricke)

Right: Norway, after she was registered in Nassau, showing the extra deck that had been constructed to provide more accommodation space. She is seen arriving at Le Havre but who can tell what the French thought of the emasculation of their graceful former national flagship. (Fabien Trehet)

Left: Norway resplendent in her final livery. (Dr Allan Ryszka-Onions)

Below: Norway, now renamed *Blue Lady*, under tow off Cape Town on her final voyage to the breakers. Once a thriving cruise vessel, now a dead ship. (Ian Shiffman)

Opposite above: *QE2* in the Bethlehem Steel drydock at New Jersey in 1977. (Bill Miller)

Opposite below: QE2 in her initial guise in Southampton Water outward bound for New York with Red Funnel's tug *Hamtun* in attendance. (Mike Lindsay)

Queen Elizabeth 2
CUNARD
BETHLEHEM STEEL CORP.

Queen Elizabeth 2
CUNARD

Above left: In her post-Falklands grey livery with Cunard funnel colours restored. Her predecessors had been painted grey to go to war. *QE2* is escorted in Southampton Water by Red Funnel's *Clausentum* and Alexander Towing's *Ventnor*. To the left of the picture is Red Funnel's car ferry *Netley Castle*. (Mike Lindsay)

Above right: QE2 while being painted with the post-Falkland's War grey hull. She is seen with scaffolding up around the after decks as she receives the finishing touches to the paint job. (Eric Brett)

Below: It wasn't long before Cunard decided to reinstate *QE2*'s original livery. Here she is seen alongside the Queen Elizabeth II Terminal taking on fuel. (Mike Lindsay)

Showing her stern pulling down in the water, *QE2* shows the power that her new electric motors delivered at speed. At this time she was still registered at Southampton. (Cunard)

Under way for another cruise, *QE2* departs Southampton on 16 April 2005. (David L. Williams)

Queen Elizabeth 2 sailing from Southampton on 14 April 1999, the 30th anniversary of her entry into service on the transatlantic run. The tug *Redbridge*, in attendance, is saluting her in the time-honoured tradition. (David F. Hutchings)

Stephen Payne's masterpiece *Queen Mary 2* departs Saint-Nazaire for Southampton on 23 December 2003, following acceptance trials, during which her future Master, Commodore Ronald Warwick, took command. (Alstom Marine)

Amid the hustle and bustle of pleasure and fishing boats *QM2* arrives at Southampton for the first time on 26 December 2003. She made a double pass as she proceeded to the Western Docks before turning and making her way to the Queen Elizabeth II passenger terminal. (David L. Williams)

Alongside at the Queen Elizabeth II Terminal on 7 January 2004, *QM2* awaits the official transfer of flagship status of the Cunard fleet from *QE2*. (David L. Williams)

A later view of *Queen Mary 2* looking pristine as she sails from Hamburg on 13 August 2011. (Martin Rulsch)

Another stern view of the majestic *Queen Mary 2*. If she is to be the last of the great transatlantic super-liners, which is certainly not a definite expectation, she has made the most impressive swan song. (Authors' collection)

SIX

STYLE AND SPEED, THE LAST GREYHOUND: THE UNRIVALLED *UNITED STATES*

When looking back at the North Atlantic passenger ships owned or built by the United States, it did not match those of British and European output. Back at the end of the nineteenth century and into the Edwardian era, American Line's *St. Paul* and *St. Louis* were the crack US liners of their day but they were comparatively small and did not amount to rivals of Cunard's and White Star's trios. Then again, people were primarily emigrating from Britain, Ireland and Europe. In the first fourteen years of the twentieth century, 13 million crossed to the US. Essentially, it was all one-way traffic, so the attitude appears to have been to let the passengers cross in British and European carriers.

After the First World War, the US had a small fleet of confiscated ex-German passenger liners such as the Irish-built *George Washington* and *Amerika* and the German-built super-liner *Vaterland*, which became *Leviathan* (see Chapter 1).

It could be said that America had no need for an indigenous-built super-liner. They certainly had the technology, the shipyards, the manpower, the expertise and the 'know-how', to which the large battleships and cruisers built for the US Navy bear witness. Post-war, the main Atlantic shipping organisation was the US Shipping Board, later the US Maritime Commission, which engaged the United States Lines to manage *Leviathan* on a 'one-horse' service with no suitable consort. There was talk and plans for a partner super-liner with some heavyweight proponents behind these intentions but nothing transpired.

In 1932 and 1933 two intermediate passenger liners, *Manhattan* and *Washington*, were built for the company and they were the first North Atlantic trade ships constructed by an American shipyard in thirty-five years. Three quarters of their construction costs were met by the US government as they were built with a view to conversion to naval auxiliaries if required. However, they were not up to the opulence and majesty of *Leviathan*. Claimed as 'the largest cabin-class liners on the Atlantic', they were instrumental in provoking a change through the North Atlantic Conference that led to the redesignation of the first-class accommodation on the much larger liners to cabin class.

In 1938, as mentioned in the Introduction, George Gaede presented his technical paper 'American Super-Liners – They Will Pay!' to the Society of Naval Architects & Marine Engineers (SNAME). In this he aired caution from 'the *Leviathan* Experience', as he called it. Based on sound and in-depth analysis of passenger travelling statistics and trends, he argued that a pair of US-built and -operated super-liners was a viable proposition. His proposals interested some serious backers such as Theodore Ferris and the eminent naval architect William Francis Gibbs. Despite his expressed view 'They Will Pay!', it was more a case of 'who will pay for their construction?'.

In the meantime Gibbs, of Gibbs & Cox, had been designing a smaller liner for the United States Lines, which entered service in 1939 as *America*. At this stage it was seriously believed that the Atlantic run could be operated viably by US ships. *America* was a twin-screw ship, not large at 33,532 gross tons and 723ft in length. Gibbs may well have observed the previous European-built record breakers like *Bremen*, *Europa*, *Rex*, *Conte di Savoia* and *Normandie*, all

United States' christening ceremony, Newport News on 23 June 1951. (Newport News SB & DD Co.)

of which favoured the bulbous bow configuration for speed and sea-keeping qualities. To this end *America*'s hull was of fine form (with a block coefficient of 0.586). Her bow was a slight clipper shape with a small bulbous forefoot. She in turn was the largest liner constructed in the US but her entry into regular service was interrupted by the Second World War, during which she served as USS *West Point*.

In 1946, *West Point* was converted back to *America* and soon established herself in the post-war Atlantic service for which she had been designed originally. She was a precursor of greater things to come, as will be seen. Surprisingly, she became one of the popular passenger carriers of the period because of her excellent cabin-class accommodation and American cuisine (as compared with the diet of families in post-war Britain and Europe because of the rationing backwash). In addition to this, it was a treat for American travellers of long memory to be able to sail on a ship with liquor freely available. *America* was no super-liner and was no ocean greyhound but she could crack along at 24 knots and had a reputation for punctuality and efficiency of service during her eighteen years on the North Atlantic, in the course of which she made some 290 crossings and carried 476,462 passengers. She was sold in 1964 to Chandris Lines of Greece for further trading.

Following the Second World War, the United States Lines eyed the resumption of the fast passenger liners of Cunard and the French Line and also, later, of the 'phoenix' Norddeutscher Lloyd in Germany. The American company wanted to be a bigger player, in keeping with the massive US contribution to the Allied victory in the war. Coupled with this, it was the burning ambition of William F. Gibbs, and his brother Frederic, to enter the North Atlantic arena with a prestigious liner that could challenge for the speed trophy.

The volume of transatlantic traffic increased each year following the war and, in July 1950, 145,266 Americans crossed in British liners to war-torn Europe, then painfully raising itself from the debris but still a powerful sightseeing magnet.

In fact, the genesis of the super-liner concept had been on Gibbs' drawing board from 1943 to 1948. No shipping company could afford to finance such a large ship as Gibbs envisaged but he managed to convince the authorities and Department of Defense that such a prestigious vessel, if required, could be converted quickly into a troopship capable of carrying 14,000 troops and able to transit both the Panama and Suez Canals. The Gibbs & Cox concern spent £350,000 on tentative plans and tests before their final designs were submitted to and approved by defence officials. Surprisingly, the US government was prepared to subsidise the ship's construction and operation. The project was given the green light and Newport News Shipbuilding & Drydock Company was awarded the contract. The specifications were submitted by the US Maritime Commission in participation with the United States Lines and the US Navy. Gibbs & Cox designed the ship to these specifications. The vessel's total cost was $78 million, of which the United States Lines contributed $35 million, the shipping subsidy board $18 million and the US Navy $25 million.

United States floated out from drydock on christening day. Note the absence of lifeboats at that time. (Frank O. Braynard)

Stirring views of *United States* flat out on her shipyard trials. (Both United States Lines)

Designated Hull 488 when her keel was laid on 8 February 1950 in No. 10 Graving Dock at the Newport News shipyard, she became the first ship of this type in modern times to be built in a drydock instead of on a conventional slipway and, therefore, she was largely concealed from public scrutiny. Ironically, more than a hundred years prior to this, Isambard Kingdom Brunel had used the same technique when building *Great Britain* at Pattersons Yard in Bristol.

Although shrouded in secrecy at the time, as it eventually turned out there was nothing unusual or radical about her underwater hull form. The ship's steel hull was the most completely welded of a passenger ship of the period, with some 1,500 miles of welding in her, and in total her construction required 1.2 million blueprints. The greater part of the hull was prefabricated in shops for assembly in the drydock in sections, more than 2,200 prefabricated assemblies in total, a large proportion of which weighed between 35 and 60 tons, while many of the largest were as heavy as 100 tons. Radiographic inspection of the welding was carried out at all stages of construction. The hull was of battleship-grade 2in-thick steel plate, welded rather than traditionally riveted. A double hull contained wing tanks for counter-flooding should the need arise and the hull was required to withstand the great storms of the North Atlantic as well as potential torpedo hits during wartime.

In place of steel, some 2,000 tons of aluminium alloy were used in the superstructure, mast, funnels and lifeboats. Unlike the hull, these sections were mainly riveted together. However, the superstructure was stiffened by steel and the shipbuilders had to adopt new techniques of riveting aluminium alloy superstructure to the steel decks and the hull below. Innovative methods for making, annealing and inserting aluminium alloy rivets were devised and also a type of insulation in the form of DuPont PAW tape ('neoprene') gasket between the steel and alloy to prevent galvanic corrosion of the two dissimilar metals in contact. (Towards the end of the decade, Harland & Wolff of Belfast would use the same materials combination and joining techniques on the P&O Line's *Canberra*).

United States at Southampton with tugs, berthing in the Western Docks. (Authors' collection)

The use of aluminium alloy significantly lowered the ship's centre of gravity, resulting in better stability, with her transverse GM believed to be around 7ft. (GM is the metacentric height, a measure of a ship's initial stability. It is the distance between G and M, where G is the transverse centre of gravity of the ship and M is the metacentre, an invisible point determined by the ship's heeling). The hull shape was of fine form with a tiny, almost insignificant, bulbous bow but a more conventional raked bow and knife edge, resulting in an underwater block coefficient of 0.559. Gibbs had been impressed by the two *Queens*' performance during the Second World War, so in that regard had Gibbs & Cox taken a leaf from the 1936-completed *Queen Mary*, which still held the speed record? The new liner's draught was 31ft, compared with *America*'s 33ft.

United States at speed in the English Channel. (Nigel Scutt/Fotoflite)

There was nothing unusual about the power plant except that, at the time, the engines were the most powerful installed in a merchant vessel. The main propulsion machinery comprised four Westinghouse Electrical Corporation of Pittsburgh, double-reduction gear, impulse-reaction turbine sets driving four propellers. Each turbine set could develop 60,500 shp, totalling 242,000 shp for a maximum designed speed of 38 knots! Steam was raised by eight Babcock & Wilcox water-tube boilers at a pressure of 990 psi, with a superheat of 900°F (527°C). For average steaming, six of the eight boilers were on stream. The power plant layout bore a strong similarity to the conventionally powered USS *Midway*-class aircraft carrier of that period with similar boilers to those used in the USS *Iowa*-class battleships.

The engine and boiler room layout comprised two boiler rooms, two engine rooms and two auxiliary machinery rooms. The forward boiler room contained four boilers, each with ten burners and each capable of raising 269,000lb of steam per hour. Situated in the same room were the feed pumps and a 1,500kW turbo-generator. Immediately aft of this boiler room was the forward turbine room, which also contained two more turbo-generators and electrical switchboards. Further aft of this turbine room was an auxiliary room with four air conditioning units and three evaporators that were capable of producing 250 tons of fresh water per day. That group formed a unit in its own right that drove the forward outboard propellers. A duplicate set of three rooms, separated from the forward set by a deep tank, drove the after inboard screws. This layout allowed each set of rooms to be operated in isolation in the event of damage disabling either group, permitting the ship to run on two screws if necessary. Equally, a complex system of pipe crossovers allowed any combination of the boilers and turbines to be cross-connected and operated if needed. Adequate 'redundancy', as the US Navy would put it.

Her four manganese bronze screws were initially to have been four bladed, but Gibbs & Cox's hydrodynamics specialist Elaine S. Kaplan had carried out extensive tests and two 18ft diameter four-bladed and two 18ft diameter five-bladed were selected, each weighing

27 tons. This configuration was tested and chosen for its superior hydrodynamic performance, in which both cavitation and resultant ship-wide vibration were expected to be greatly reduced. The forward outboard, four-bladed screws rotated at 139 rev/min and the after inner pair ran at 141 rev/min, which gave the ship a service speed of 29.5 knots. This incremental speed difference between the two propeller pairs was also intended to reduce vibration.

On 23 June 1951 the new liner was resting on her blocks two thirds finished, with engines, boilers, bridge and funnels in place and her underwater hull form already immersed in the water. She was christened *United States* by Mrs Tom Connally. Unlike a conventional launching of a ship sliding down the ways into the water, the drydock, which had been partially flooded the previous night, was fully flooded and *United States* floated off her keel blocks. There was still plenty of fitting out to do and Gibbs & Cox ensured that every technological advance of the day that could be afforded by the owners and the US Navy would be fitted into the sleek hull. She was fully air conditioned and open decks throughout were covered by a fire-retardant composition called 'Neotex' instead of teak. This decking had to be pleasant to walk on and one that would not crack or warp or become slippery in wet weather. To this end, for six months a section of 'Neotex' was placed on a road serving the shipyard used daily by thousands of shipyard workers and driven over by hundreds of heavy trucks before the material was accepted and enough ordered to cover a 60,000 sq ft area of deck. Some 500,000 sq ft of fire-retardant 'Marinite' asbestos material was used throughout in the panelling, insulation, floor tiles and pipe lagging. During the construction as many as 3,500 workmen were busy in the shipyard at one time.

United States emerged as the largest passenger ship built in the US. She measured 53,329 gross tons, 990ft length overall, 101ft 6in width and 175ft from the keel to the top of her foremost funnel. She had twelve decks that catered for 910 first-, 560 cabin- and 537 tourist-class passengers with a crew of 1,000. At the time of her completion she was the third largest liner in the world after the two *Queens*. It was rumoured that each of the then forty-eight states and all of the US territories had contributed in some way to the construction of the ship as well as her décor and appointments, the latter of which were described by some as 'minimalist'.

Her first acceptance trials began on 14 May 1952 off Virginia with 1,700 people on board and during this she was reputed to have averaged 38.32 knots sailing into a 20-knot headwind, with a speed of 42 knots achieved at full power. Her sea-keeping qualities came to the fore in rough weather and for stabilisation, the long, balsa wood-filled bilge keels were considered sufficient to minimise rolling as no fin stabilisers were fitted. In addition, the weight-saving aluminium alloy superstructure reduced the radius of gyration (gyradius), making the ship more sensitive to sea-induced roll motions. The ship's beam of 101ft 6in was determined by both the need to fit the high-power steam machinery and satisfy troop transport stability needs, besides the ability to transit the Panama Canal. Her long length of 990ft, wide beam and low gyradius could have resulted in severe motions. Unlike her contemporaries, *United States* had huge funnels that not only supplied an increase in gyradius but had an obvious above-water stabilising fin shape, streamlined air foil profile facing ahead but presenting large flat, high-drag plates to the side.

She was ready for commercial service and, fortuitously for the United States Lines, had just missed being taken up for use as a troopship while the Korean War was still raging. Whichever, this 'greyhound' was straining at the leash!

On 3 July 1952, *United States* commenced her maiden voyage under the command of Commodore Harry Manning, with her designer on board for the first and only time. Despite gale force winds, on the eastbound voyage she averaged 35.59 knots and covered the 2,949 nautical miles between Ambrose Light and Bishop Rock in three days ten hours and forty minutes, beating *Queen Mary*'s 1938 record by some ten hours over the same distance. Upon her arrival, a huge flotilla of welcoming craft, mainly yachts, tugs and tenders accompanied the liner along Southampton Water, while an estimated 70,000 people turned out to watch her arrival at Southampton's Ocean Terminal. United States Lines officials, passengers and crew on board were overwhelmed by the warmth of the greeting as it had been thought the British might be a bit stand-offish, especially as the Blue Riband was being snatched away from *Queen Mary*. Commodore Harry Manning touched on this point when he replied to the official welcome given on board the liner: 'I'm very happy to have won the Blue Riband, yet there is a tinge of sadness in taking it away from such a fine ship as the *Queen Mary*, with years of excellent service and the fine traditions of the British merchant navy behind her.'

One message of congratulation Commodore Manning received came from the then prime minister, Sir Winston Churchill.

On the return leg of her maiden voyage, on 11 July, she steamed 2,906 miles between Bishop Rock and Ambrose Light in three days twelve hours and twelve minutes, once again beating *Queen Mary*'s record by nearly ten hours with a mean speed of 34.51 knots. She had captured the Blue Riband east- and westbound and was awarded the Hales Trophy, which for many years was held at the United States Merchant Marine Academy at Kings Point, New York.

United States arriving alongside at Southampton's Ocean Terminal on 8 July 1952 on the completion of her record-breaking maiden voyage. (SuperStock / Alamy Stock Photo)

Under normal service conditions *United States* maintained a sustained speed of 28.30 knots with a daily fuel consumption of 700–740 tons, as opposed to 850 tons per day at full speed. This gave her a range of 10,000 nautical miles without refuelling. She required forty-seven engine room staff on a three-watch system. In November 1952, she was drydocked for hull examination and to have her propellers changed.

There had been an increasing upsurge in people travelling by air, for by 1948 252,000 passengers preferred to fly rather than go by sea and by that time the ships had lost a third of their pre-war passenger traffic to the airlines. In 1951 a total of 167,160 Americans visited Britain, 43 per cent of whom travelled by British or American airlines, figures that continued to increase sharply. By mid-1952 passenger traffic by air had increased a further 12 per cent, and in one month alone, in October 1953, airlines arriving in the UK carried a record number, 7 per cent higher than the previous twelve months. In 1953 *United States* along with fifty-four other liners worked the North Atlantic run and in that year these ships carried a total of 837,774 passengers, 491,835 of them westbound and 345,939 eastbound, the largest number since 1930. But the writing was on the wall.

United States ran in consort with the smaller *America*, with their European terminus extended to Bremerhaven during the winter months, and both liners offered serious competition to British and European passenger liners throughout the 1950s and '60s. Immediately acquiring the sobriquets 'the Big Ship' and the 'Big U', during her first year of operation *United States* carried some 70,514 passengers.

Fifteen months later, on 21 October 1953, William Francis Gibbs was awarded the Franklin Medal at a dinner in Philadelphia. In a speech he referred to *United States*' record-breaking maiden voyage and pointed out that her propeller shafts: 'have never stopped in the 15 months by reason of mechanical derangement nor has the ship been a minute late sailing because of a mechanical defect'.

By contrast, many years later, in August 2007, Professor Jacques Hadler, former head of hydrodynamics at the David W. Taylor Model Basin at Bethesda, Maryland, later the former Dean of the

On the same occasion, *United States* at her Ocean Terminal berth flying bunting in celebration of her speed record. (Phil Fricker)

In drydock at the Norfolk Navy Yard on 28 November 1952, *United States* having her propellers changed. (Operation 2022 / Alamy Stock Photo)

Webb Institute and also considered America's expert on conventional propeller design, revealed:

> The propellers did their job on the record-breaking trip but when the ship came back into port, they had severe propeller damage. One was so badly eroded at the root you could see through 8 or 9 inches. As a result, the propellers ended up being designed three times to address two problems: cavitation erosion and vibration. More blades can deliver more power, but fewer blades are more efficient and create less turbulence. It's something we in later years, by the 60s, got a handle on.

Hadler had started his career in the US Navy working on the speed trials for *United States* and he also worked for the men who redesigned her propellers.

Once settled into service, *United States* soon attracted popularity from the great and the good, royalty, film stars and politicians alike, such as Marlon Brando, Elizabeth Taylor, Sean Connery, Grace Kelly, Prince Rainier of Monaco, Walt Disney, Salvador Dali, four US Presidents, the Duke and Duchess of Windsor and John Wayne.

The year 1957 was a tipping point in the transition of North Atlantic traffic when the air/sea balance was reached for the first time, there being 1,036,000 passengers who crossed by sea and 1,032,000 by air. After this the proportion crossing by sea declined steadily. If one desired to cross by sea, the fastest passage was, of course, on *United States*, with a more regular crossing time of just over four days. Alternatively, if one wanted a faster crossing from London to New York, a six-and-a-half-hour flight with TWA was an option. But what would one do with the time saved?

It may have been his visit on board the liner that was the genesis for Walt Disney to produce the 1962 film *Bon Voyage* starring Fred MacMurray and Jane Wyman; a light-hearted comedy about a family trip from New York to Paris travelling aboard *United States* with some location shots taken aboard her.

With *United States* having been in service for over a decade she remained largely unaltered structurally and she continued to maintain a regular punctual service from New York to Southampton and Le Havre and from Southampton, returning via the same ports. On 7 March 1963 the liner departed New York carrying an extremely valuable treasure. Leonardo da Vinci's painting the *Mona Lisa* had been on loan to the US from the Louvre, Paris, for eleven weeks, first on exhibition in Washington D.C. and then at the New York Metropolitan Museum of Art. A suite of cabins was allocated and the painting was lodged in a temperature-controlled room with a twenty-four-hour armed guard mounted (although if the painting had been the victim of theft, one wonders what any prospective perpetrator was going to do with it in mid-Atlantic!). The 'Mona Lisa' was landed at Le Havre and safely restored back in the Louvre on 12 March.

The already-mentioned *America* made her last Atlantic voyage in November 1964 and was sold on to the Chandris Group for further trading as *Australis*. This left *United States* without a running consort, although she was co-scheduled for a time with *France* (see next chapter). Even so, with the increasing inroads made by air travel, it was hard for her to maintain a viable service. In addition to this, between 1961 and 1969 the US government's operating subsidy for *United States* had amounted to $72 million. In an attempt to reduce the berthing and pilotage charges and drydock duties, her tonnage was 're-measured' at 38,216 during 1967. This effectively, though questionably, relegated her from the super-liner league. Sadly, during that same year her designer, William Francis Gibbs, died.

During February 1969 she undertook a cruise to South Africa, a most unusual choice of itinerary for Americans but an opportunity for them to call at another country as opposed to those across the North Atlantic.

Unbeknownst to her passengers and crew and the public, upon her arrival at New York, *United States* completed her last voyage (No. 400) on 7 November 1969 after seventeen years on the Atlantic. Ironically, in late 1969, before the decision was made to retire her,

Above: United States eased out of the Ocean Terminal assisted by the original tug tender *Calshot.* (B.L. Devenish-Meares)

Left: In Southampton Water at the end of an eastbound crossing, *United States* with Alexandra Towing tugs. (Phil Fricker)

United States in the floating dock *Norshipco Titan* during her final refit at the time when her service life was abruptly ended. (Norfolk Shipbuilding Corp)

the United States Lines announced a fifty-five-day Grand Pacific Cruise, which was set to commence on 21 January 1970 but this was cancelled in December 1969.

After her final arrival at New York, *United States* sailed to Newport News for her scheduled annual overhaul. While there, the United States Lines announced its decision to withdraw her from service. The decision was due to the skyrocketing expenses of operating the vessel plus a combination of competition from the airlines, union intransigence and principally the US government's discontinuation of its operating subsidies. The news stunned her officers and crew, some of whom had left their equipment and personal items on board in anticipation of returning to work the following day, only to find the gangway doors sealed shut. The announcement halted all work on the ship, with many maintenance tasks left incomplete. The ship was sealed up, with all her furniture, fittings and crew uniforms left in place. Essentially, *United States* had reverted back into the US Navy as a reserve ship.

In June 1970 she was relocated across the James River to the Norfolk International Terminal in Norfolk, Virginia. By 1978 the US Navy felt that *United States* was obsolete as far as naval use was concerned and the ship was no longer relevant to their needs. Returned to the US Maritime Administration, successor to the US Maritime Commission, she was put up for sale.

The ship passed between various owners with projects for her future beset by constant false starts. In 1984, in order to pay accrued debts, the ship's fittings and furniture, which had remained entombed in her since 1969, were sold off at an auction in Norfolk, Virginia. The auction, which lasted from 8 to 14 October 1984, raised $1.65 million from 3,000 bidders for various objects from the liner. Some of the artwork and furniture went to various museums, including the Mariners' Museum of Newport News.

Later again, on 4 March 1989, after languishing for so long, she was towed across the Hampton Roads to the CSX coal pier at Newport News. A rather bizarre catalogue of events was to follow when the ship was seized by US marshals for non-payment

Half-painted – see her funnels – she is towed away by tugs, her refit suspended incomplete following the termination of her operating subsidies. (Norfolk Shipbuilding Corp)

United States with tugs in Turkey at the time when asbestos and other fittings were removed from her. (United States Conservancy)

of mortgage and docking fees and put up for auction by the US Maritime Administration on 27 April 1992. The ship was then bought for $2.6 million by a group called Mamara Marine Inc. with Turkish interests.

On 4 June 1992 the ship left Newport News bound for Istanbul under tow of the tug *Smit Rotterdam*, reaching the Sea of Marmara on 9 July and subsequently laid up at Tuzla. It was intended that she be refitted as a cruise ship and she was registered in Panama during 1993. On 1 November 1993 she was towed to the Sebastopol Shipyard in Ukraine. While there, she underwent asbestos removal in which the interior of the ship was almost completely stripped down to the bulkheads. Her open lifeboats, which would no longer have met the new SOLAS requirements, that is if the ship were to sail again, were also removed and scrapped along with their davits. In this time Mamara Marine ran into financial difficulties and on 15 May 1994 the ship was returned to Tuzla. Once again, on 25 July 1996, the old *United States* was towed back to the Philadelphia Naval Dockyard by the tug *Smit New York*. The total cost of towing her to Ukraine and back had amounted to $50 million!

While at Philadelphia, a Mr Edward Cantor purchased the ship in November 1997 for $6 million and two years later the SS *United States* Foundation and the SS *United States* Conservancy succeeded in having the ship placed on the US National Register of Historic Places.

At the time of writing the sleeping giant was still laid up at Pier 82 on the Delaware River at Philadelphia and it would be wrong to speculate on her future.

United States' record stands:

The first super-liner to be built after the Second World War.
The largest passenger ship to be built in the United States.
The first ocean liner to make extensive use of aluminium alloy.
The fastest ocean liner ever built.
One of the few passenger liners built to transit the Panama Canal but which never actually made the transition.

In her time *United States* had completed 400 Atlantic crossings, steamed some 772,840 miles and carried some 1,025,700 passengers. Many would say she was the last 'true' holder of the Blue Riband of the Atlantic Ocean – indeed, the Last Greyhound!

SEVEN

REBIRTH OF A LEGEND OR UNIQUE SUPER-LINER EXTRAORDINAIRE? THE CGT'S IMPRESSIVE *FRANCE*

Despite having invested heavily in the restoration of *Liberté*, the French Line (CGT) continued to aspire for a more fitting replacement of *Normandie*. From a practical point of view, orders for replacement ships had to be considered urgently because both of CGT's frontline ships, *Ile-de-France* and *Liberté*, were ageing and would reach retirement age by 1959. Successors could take three to four years to build, necessitating an order by 1955 latest. Equally, CGT was aware that its competitors were contemplating upgrades or additions to their premier ships, Cunard developing its replacement strategy for the two *Queens* while a possible running mate for *United States* was also on the cards.

Hence, barely two years after *Liberté* had commenced operations for the company, rumours began to circulate in the trade press, as early as the autumn of 1952, that the French Line was planning a new super-liner. The company's design office drew up outline plans for a new transatlantic liner and on 16 November of the following year, a contract was placed with the Penhöet shipyard, Chantiers de l'Atlantique since 1955, to explore two options: a five-day, 31-knot ship capable of carrying 2,000 passengers and a six-day, 25-knot liner with the same passenger capacity. Apparently, a third option for two smaller vessels of around 35,000 gross tons for the New York run had been considered but rejected in favour of a single giant ship dedicated to all-year-round transatlantic service.

It came as no surprise, therefore, when in March 1955 there was an announcement that in principle a 60,000 gross ton, 30-knot ship was planned for the Atlantic service that would replace both *Ile-de-France* and *Liberté*. She would carry the name *France*, in honour of the nation and as a tribute to two previous CGT ships of that name. But a lot of water was to go under the bridge before the point in the proceedings was reached where an order could be placed.

Talk began immediately of the prospect of a virtual clone of *Normandie* materialising, creating intense excitement among the French population. Such speculation was inevitable. Given that the new liner would be of comparable size, and no doubt equally glamorous, could it possibly be a kind of doppelgänger of the fated earlier vessel? The conjecture intensified later when, after his wife Madame Yvonne de Gaulle had christened the then incomplete *France* at her launch on 11 May 1960, President Charles de Gaulle declared jubilantly, 'I have given you a new *Normandie*.' Early artistic renditions of *France*, before her distinctive winged funnels had been decided upon, also tended to suggest certain visual similarities.

The reality was though, that *France* was a different ship for a different time and as such she stood out in her own right as a magnificent ocean liner of individual style, character and Gallic flair, as well as being extraordinarily elegant to behold given her size. Of course, de Gaulle had been deliberately appealing to the nostalgia and patriotic sentiments of the French people at a time when the country was still recovering from the devastation and divisions inflicted during the recent war as well as reeling from the loss of colonies in the Far East and North Africa.

On the slipway at Saint-Nazaire, the bow section of *France* being fitted on 24 November 1959. (Ch. de l'Atlantique)

Considering the persona *France* was expected to inherit, some comparisons with *Normandie* should be made, if only to demonstrate the differences between the two vessels. From a visual standpoint, *France*'s long, low hull revealed a curved, almost clipper-like stem that had a more elegant and more inclined arc to it than that of *Normandie*, while at her stern end she had a rounded cruiser-spoon stern instead of the elliptical counter of the earlier ship. Of lower tonnage than *Normandie*'s 83,423 gross, *France* measured 66,343 gross tons as built, which accounted for her having a generally sleeker and leaner look than her forebear. Her bulbous forefoot projected further forward. She retained the foredeck turtle-back for water dispersal when shipping it green (when large sea waves break over a ship's bow and onto its foredeck) but had derricks with fore and aft spars, whereas *Normandie* had two cranes mounted just ahead of her bridge front. Similarly, *France* had a modern, broad aluminium signal mast, while *Normandie* retained two wooden masts that, although considerably shorter and slightly inclined befitting her streamlined profile, were still of more traditional design. And of significance, *France*'s winged funnels, just two rather than three, completely set her apart visually. The fact was though that both ships in their own way had strikingly attractive lines but they came from different moulds.

The design of *France* developed rapidly enough through an intensive series of trials and tests in the model testing tanks in Paris but the financial aspects of the project's gestation dragged on interminably. Discussions with the French government had commenced in November 1953. However, with acrimonious debate, political haggling, diversionary tactics and foot dragging in the corridors of power in the French parliament, it seemed there would be no resolution to the matter of funding between the government, the shipyard and the French Line regarding the overall cost and its distribution between the parties, the schedule for stage payments and so on.

Not everyone in the French parliament shared de Gaulle's enthusiasm for the great liner project. It was noted that earlier, de Gaulle had to some extent been influential in dissuading the French Line from ordering the smaller 35,000 gross ton ships in favour of the single giant. Given that the construction of *France* would depend on finance from the public purse besides open-ended state subsidies throughout its working lifetime, he had determined that, even so, the state should make such a bold commitment on the basis that it would be better for French national pride to have a single, awe-inspiring super-liner than two nondescript intermediate ships. Thus, it was clear that he saw *France* as a statement of national prestige regardless of the cost, rather than, necessarily, a viable commercial proposition. Not that this in any way detracted from *France*, which proved herself to be immensely popular, a marvel of design and a symbol of continued grandeur in a changing world.

Briefly, as the political shenanigans persisted, the concept of the two 35,000 gross ton sisters resurfaced, as well as a suggestion that the 16-year-old liner *Pasteur*, built originally for the River Plate service of Compagnie de Navigation Sud-Atlantique and surplus to requirements after a spell as a troopship, should be renovated for the French Line's North Atlantic requirements to serve along with a new 1,000 passenger vessel of around 30,000 gross tons. One by one, these distractions were swept aside and at last, after three and a half years, a settlement was reached, but the lengthy delay had set back the new ship programme by two years and required the retention of *Liberté* until the end of 1961.

The launching of *France* on 11 May 1960. (Ch. de l'Atlantique)

Bow view of *France* after the launch as she is manoeuvred into the Penhöet Dock for fitting out. (Authors' collection)

At the heart of the procrastination was the prickly subject of the construction funds, along with the pros and cons of the two concepts developed by Penhöet that, for different reasons, had their proponents and opponents. A study revealed that the five-day, 31-knot vessel would have capacity for some 82,600 passengers per year, whereas the six-day, 25-knot variant could only attract 63,000 passengers per year. This revelation finally settled that particular issue. As for the financial aspects, though, they were a lot more involved. The construction of *France* would benefit from construction subsidy legislation introduced in the Shipbuilding Act of 24 May 1951. This would provide a contribution of 7 billion francs to the cost but the French Line required greater financial support. Penhöet's initial estimated building price was 31.75 billion francs, towards which the French Line had a budget of 17.5 billion francs, leaving a shortfall of 7.25 billion. To break the impasse, CGT accepted the state funding as determined by the framework of the Shipbuilding Act and in continued negotiations with the shipyard reached the lower construction cost of 27.36 billion francs. The subsidy was raised to 7.66 billion francs and the French Line increased its contribution to 19.7 billion francs. The sums finally added up.*

Although the fiscal commitment was finally agreed upon, the legislation to permit the release of credits was not passed until 25 July 1957, a year to the day after the official contract to build the *France* was signed between Jean Marie, the CGT chairman, and René Fould, the chairman of Chantiers de l'Atlantique, at the French Line headquarters in Paris on 25 July 1956.

There remained some disquiet about committing to a ship designed for year-round transatlantic service, a strategy viewed in some quarters as risky. Hardly had the keel of the new ship been laid when industry statistics clearly revealed that ocean liner travel and airline travel were on opposing trajectories, especially for the premium first-class traffic. Although the number of passengers travelling by ship remained high, higher in fact than ever before at over 1 million in 1956, in short order the number declined and was soon overtaken by passengers numbers on flights. And the gap continued to widen. By 1960, twice as many passengers crossed the Atlantic by air than by sea. Three years later, with *France* in only her second season of scheduled sailings, the liners on the Atlantic run for the first time carried fewer passengers than they had the previous year. The very essence of what the super-liners had been conceived to provide – faster crossing times in luxurious surroundings – was threatened. It was becoming increasingly anachronistic. Instead of four or five expensive days at sea, however good that once was, travellers could now cross the Atlantic in a matter of hours for relatively modest fares, and that had unchallengeable appeal. Nonetheless, six years earlier, as construction of *France* was getting under way, the French Line pressed on regardless with its dedicated transatlantic super-liner.

France alongside the fitting-out quay before her funnels had been installed. (Ch. de l'Atlantique)

* It should be noted that these figures are expressed in 'old' French francs. From 1960, to deal with progressive diminution of the franc's value, the 'new' French franc was introduced in which each 'new' franc had the worth of 100 'old' francs.

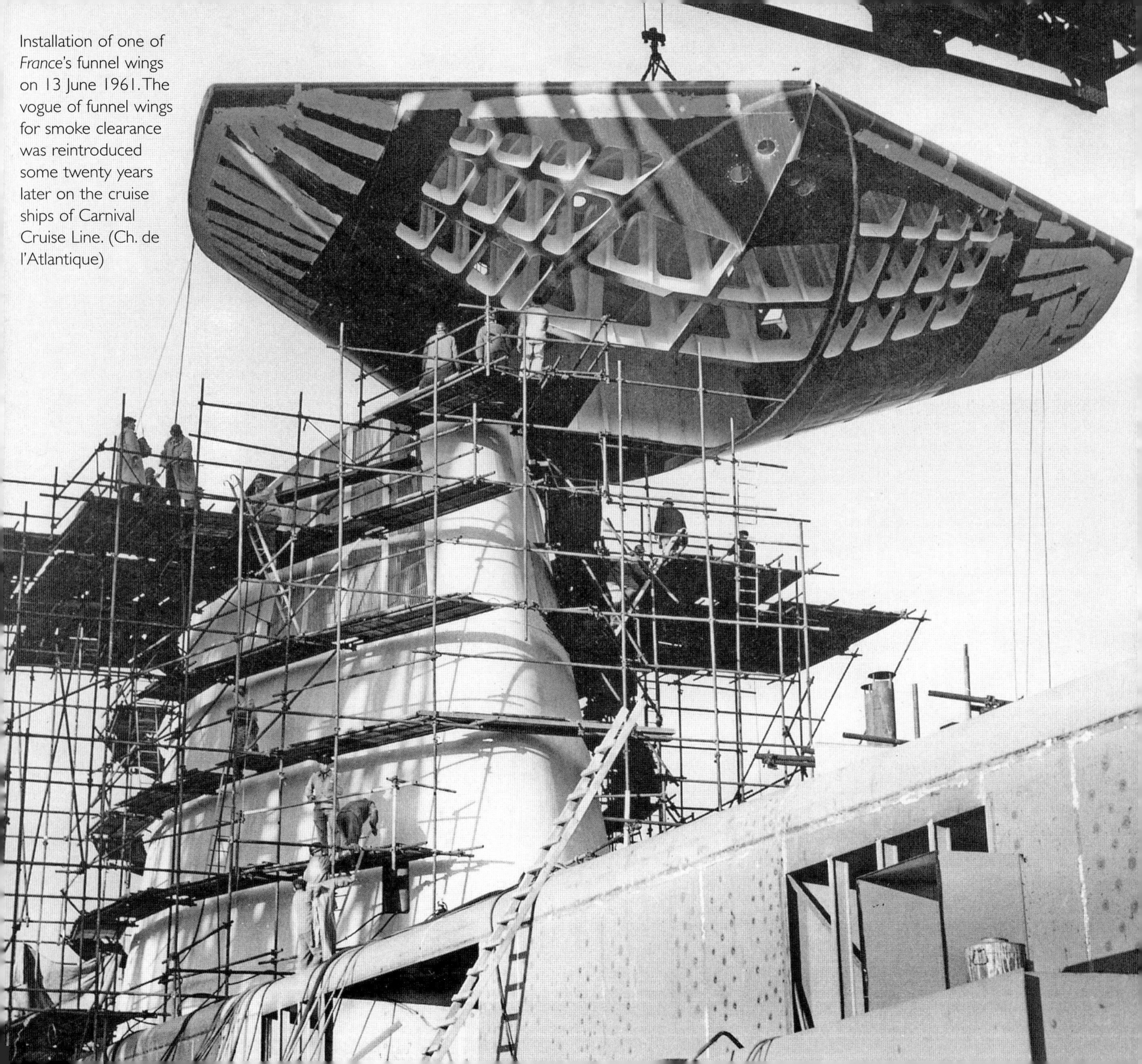

Installation of one of *France*'s funnel wings on 13 June 1961. The vogue of funnel wings for smoke clearance was reintroduced some twenty years later on the cruise ships of Carnival Cruise Line. (Ch. de l'Atlantique)

Construction of Yard No. G19, as she was designated, commenced on 7 October 1957 with the laying of the first keel plates. The method of her assembly did not follow the traditional ribs, beams and plates system that was still generally employed by shipbuilders at that date. Nor was it entirely a modular system of construction, although some hull and super-structure sub-sections were constructed elsewhere in France and transferred to the shipyard, where they were welded together. As revealed by sequential photographs, the growth of the hull and subsequently the super-structure appears to have followed a deck-by-deck approach, whereby deck and hull plating was progressively welded together at each level. By this method of assembly, *France* became the first of the super-liners to be fully welded. *France* would displace just 51,815 tons, whereas *Normandie* had weighed in at 68,350 tons, the use of aluminium in the fabrication of the superstructure from the verandah deck upwards accounting for this significant weight difference.

Work continued apace and two years and eight months later she was launched. Subsequently, in the Joubert drydock, her four propellers and her massive 74-ton rudder were installed and then, once refloated, she was moved to the fitting-out quay in the Penhöet Dock, where her upper structure was completed, her distinctive funnels were placed in position and the myriad tasks undertaken to prepare her interior accommodation spaces: the fittings, the lighting, the decorations, everything befitting the luxury, grandeur and refined ambience intended for this new national flagship.

When *France* was virtually complete, on 11 and 12 November 1961 the Penhöet yard workers and their families, more than 8,000 in total, were invited to visit her and see for themselves the result of all their labours, the wonder of their creation. Initial sea trials followed off Belle-Île in the Bay of Biscay, commencing on 19 November 1961 and lasting for three days, with surprising results. *France* had been designed to be a fast ship but the sustained achievement of just under 35 knots for some hours over a considerable distance exceeded expectations. It was thought she might be capable of wresting the Atlantic Blue Riband from

Complete and ready to leave the fitting-out quay for her sea trials, the gleaming new *France* in November 1961. (Jean-Yves Brouard collection)

France arrives from her builders and makes her first entrance into the Port of Le Havre. (Kenneth Wightman)

United States but, in reality, the American ship had a lot in reserve over her 35.59 knots average record speed.

The power, propulsion and electricity-generating machinery of *France* comprised four sets of Compagnie Electro Mécanique (CEM)-Parsons single-reduction geared turbines, built under licence, laid out in two fore and aft engine rooms and driving quadruple screws. In each case, a four-stage system was employed comprising HP, IP1, IP2 and LP turbines. There were two boiler rooms. Steam was provided by eight (4×2) high-pressure, super-heating, oil-fired, water-tube Penhöet boilers. Each boiler was capable of producing 90 tons of steam per hour at 950 psi and 930°F. The power system generated 160,000 shp for a service speed of 30 knots and 175,000 shp maximum. Fuel oil consumption at service speed was 750 tons every twenty-four hours.

On completion of the first round of trials, *France* sailed north from Saint-Nazaire to enter her home port, Le Havre, for the first time on 23 November 1961, vast crowds watching her arrival. As she proceeded to the Quai Joannès Couvert, she passed the moored and laid-up *Liberté*, awaiting her fate. The salutes between the two ships signified the transfer of flagship honours from the *Grande Dame* to the *Jeune Mademoiselle*.

Subsequently, on 16 December 1961, *France* returned to the shipyard for more sea trials off Belle-Île. All went well and, these completed successfully, she had returned to Le Havre by 6 January 1962, where the formal hand-over and acceptance ceremony took place. That same day, *France* crossed to Southampton, where on a brief call, again occasioned by thousands of well-wishers, docking rehearsals were performed at the Ocean Terminal. The French Line wanted to ensure everything involving their new flagship unfolded without a hitch. With everything finalised, the great French super-liner was ready to go.

France measured 1,036ft long overall, making her the longest liner in the world. At the same time, she was also the third largest ocean liner, a ranking that later changed to make her the largest after the retirement of Cunard's original *Queens*. She held this status until 1972, when she was overtaken by *Queen Elizabeth 2* by just 500 gross tons following modifications to the Cunarder carried out that year. *France*'s beam measurement was 110ft, giving her a fine length to breadth ratio of 1:9.4.

Unlike her predecessors, which had three grades of passenger accommodation, *France* was essentially a two-class ship, an indication of the progressive democratisation being introduced to ocean travel. Her complement makes a striking comparison with that of *Normandie*'s passenger breakdown that, favouring the top end of the spectrum, had composed 848 first, 670 tourist and 454 third class. In contrast, the accommodation division of *France* was 407 first and 1,241 tourist class, plus 145 interchangeable cabins for either 210 first- or 370 tourist-class passengers. As a consequence, her maximum attainable passenger capacity if all berths were occupied was 2,044. A concession to possible change in the future was the arrangement of *France*'s accommodation in such a way that it permitted adaptation to a single class for cruise operation, such as had been done on Holland America Line's new flagship of 1957, *Rotterdam*. Her crew numbers also varied between line voyages and cruise excursions, ranging from 1,104 to 1,253.

Having created an aura of high expectation of a ship of superlatives, there was some discontentment with the interior appointments of *France*. Making more unavoidable differentiations with *Normandie*, the decorative style of *France* was regarded as smart, comfortable and functional, yet definitely stylish, but it was also regarded as rather bland and certainly lacking the refinement and ostentation of CGT's pre-war flagship. New regulations applicable to fire prevention, particularly pertinent given the cause of *Normandie*'s demise, frustrated some of the more ambitious intentions of her designers, led by Antoine Barthélemy. Rather than the open vistas, the fine veneers and flamboyant decorative frescoes, murals and other displays, the styling aboard *France* featured a greater use of glass, chrome and modern synthetic materials.

Before entering line service, *France* made a fully booked shakedown cruise to the Canary Islands from 19 to 27 January 1962. En route, while returning to Le Havre, she passed *Liberté* for a second time, the old liner then heading south to the Mediterranean for breaking up in Italy.

France made a trial visit to Southampton on 6 January 1962 for docking tests at the Ocean Terminal. The docking fees then were £2,000 a year for twenty years to help pay for the dredging of an additional 100ft of deep water. (British Transport Commission)

Left: Low aerial view of *France* at speed. Despite the French company's seemingly successful operational experience with *Normandie*'s turbo-electric propulsion system, *France* returned to the tried and tested steam turbine drive. (Nigel Scutt/Fotoflite)

Below: France passing Manhattan in 1965. (Bill Miller collection)

Finally, on 3 February 1962 she set off on her maiden voyage to New York with 1,958 passengers, the well-wishing crowds lining the shore line to see her off reaching as far as Le Bout du Monde and Saint-Adresse. After a smooth crossing, *France* arrived at New York five days later, where she was afforded the usual welcoming reception of fire boats spraying water into the air, aircraft and helicopters flying overhead and thousands lining the Battery to greet her. After a three-day turnaround, accompanied by numerous civic and diplomatic gatherings aboard her, she set off on the return leg of her maiden voyage, arriving at Le Havre on 16 February. Her passengers on the eastbound voyage were 1,829.

Thereafter, *France*'s short career of just twelve years on the transatlantic run was largely incident free, marked only by the initial regularity of her crossings, followed by the steady decline in passenger numbers in parallel with a growing government deficit required to keep her operational. In that first season, she completed forty-four Atlantic crossings. Had that schedule level been maintained over the seasons that followed, she would have clocked up a total of 528 crossings by her retirement. As it was, her total crossings were only 377, carrying a total of 588,024 passengers, still a respectable achievement.

More and more, though, she was diverted to cruising to help balance the books, even though her design did not readily lend itself to these duties. Cruise passengers wanted open-air swimming pools, spacious open decks with lido areas during visits to Mediterranean ports and on tours that took *France* through tropical and sub-tropical climates, facilities she could not provide. Nonetheless, her excursions were constantly popular and, of all the transatlantic super-liners, with the exception of the later *Queen Elizabeth 2*, *France* clocked up the most cruises and the most cruise miles. In total, she completed ninety-three cruises, including two world cruises, carrying in excess of 113,850 passengers. Her first cruise to Rio de Janeiro was in February 1963, the first to the Caribbean in the following month and her first Mediterranean cruise in March 1966.

The cruises were invariably themed; in April 1969 it was an 'Imperial' cruise in recognition of the Emperor Napoleon, which included a call at the island of Saint Helena. Her first world cruise, from 5 January to 10 April 1972 was promoted as 'Around the World in 80 Days', taking its title from the book by Jules Verne. It took *France* around Cape Horn and the Cape of Good Hope because she was too big to pass through the Panama Canal, but at least her sturdy construction designed for the stormy conditions of the North Atlantic proved ideal for the notoriously wild seas of the Southern Ocean. *France* was the first of the transatlantic super-liners to undertake a peacetime circumnavigation. A second world cruise took place in 1974, from 4 January to 11 April, but this excursion was to end on a sombre note.

Matters came to a head for *France* in 1974 following the world oil crisis of the previous year, when the price of fuel oil quadrupled within months. Precipitated by the Organisation of Petroleum Exporting Countries (OPEC), led by Saudi Arabia, an embargo was imposed on supplies of crude to the USA and certain other countries, including the UK, in retaliation for supporting and supplying arms to Israel during the Yom Kippur War. Although the embargo was not inflicted upon France, the impact internationally was to force prices up everywhere as the world economy was affected generally. At a time when the French government was seeking to make fiscal savings as well as switch support funding to the Concorde project, the subsidies to offset the operational losses of *France* became the victim of the cuts made to public spending.

The withdrawal of the subsidy was announced in April 1974 and, during her second world cruise it was disclosed to the passengers that operation of *France* on the transatlantic run would cease that year and there would be no further cruises. An official communique from the French Line, released on 8 July 1974, provided more detail of the predicament, recognising that the situation had become untenable in the absence of clarity. It was stated that *France* would be laid up at the end of the season, on 25 October 1974. The statement also declared that there was no intention to sell her or to have her broken up.

France in her prime, passing Cowes, Isle of Wight, outbound. (Peter Bridle)

Laid up in the Maritime Canal, Le Havre in 1978, *France* at the so-called 'Quai d'Oblivion'. (Chris Howell)

Her final New York departure from Le Havre was on 30 August 1974, her return to her home port expected twelve days later on 11 September 1974, but on her arrival off Le Havre her crew mutinied in protest at the impending forced retirement, preventing *France* from entering port. She was anchored off the harbour entrance, blocking the main channel, and the passengers had to be ferried ashore on a car ferry to permit completion of disembarkation.

The strike persisted for almost a month, during which time severe weather on 24 September compelled a move to an anchorage off Saint-Vaast-la-Hougue, in the lee of the Cherbourg peninsular. Having failed to achieve their objectives – the restitution of scheduled operations plus a generous pay rise – the protest ended on 9 October 1974. Ironically, the union action, rather than preventing her retirement, had only served to bring it forward by fourteen days. *France* then entered Le Havre, first tying up at the Joannès Couvert terminal, but from 19 December that year she was moved to an ignominious backwater on the Seine Maritime Canal. Dubbed by the press as the 'quay of oblivion' she was abandoned there, abject and forgotten. Close to the refineries and chemical plants in the industrial area of the port, the forlorn vessel was blanketed by their pungent fumes.

There she remained until 26 January 1979 apart from a brief move in November 1976 to Le Havre's Graving Dock No. 7 to clear marine growth off her underwater hull. Throughout the four years and four months of her lay-up, various schemes were proposed for the idle French flagship. As something of a dichotomy – Saudi Arabia having been the catalyst of her predicament – an offer was made for *France* in 1977 by the Saudi millionaire Akram Ojjeh, whose intention was to have her converted into a floating museum of French art and heritage. On the downside, it was also proposed she would become a hotel and casino relocated to one of the US southern states on the Atlantic coast. Akram had, in fact, purchased *France* for $24 million but this did not prevent the deal from falling through.

Bids were also received from Soviet interests and, later, from a Chinese source; the first to turn her into a floating hotel on the Black Sea, the other for conversion into a floating industrial trade fair, presumably promoting Chinese products. Neither scheme materialised.

The rescue, when it came, was to reinstate her active status as a passenger ship, albeit operating as an exclusive luxury cruise vessel. Engineers working for her prospective new owner, the Norwegian shipowner Knut Kloster, inspected *France* in April 1979 and were impressed by the state of repair in which they found her. Besides the quality of her construction, she had been maintained meticulously and despite the lengthy sojourn, her engines were in excellent condition. Two months later, on 26 June 1979, she was acquired for $18 million, the intention being to have *France* converted from the world's largest passenger liner into the world's largest cruise ship, managed for Lauritz Kloster by the Norwegian Caribbean Line (NCL). She would become the first and only purpose-built transatlantic super-liner to be adapted so completely in this fashion.

Renamed *Norway*, the former *France* left Le Havre under tow of the ocean-going tugs *Abeille Provence* and *Abeille Bretagne* on 18 August 1979, bound for the Hapag-Lloyd shipyard in Bremerhaven where the conversion was to take place. Originally, the departure had been planned for 15 August to permit more of the populace to see her off as that was a public holiday. However, strikers had blocked the locks from the canal and the crews of the Abeille tugs refused to assist in the manoeuvres to get her into the main channel. The truth was that the French population were largely bereft at the loss of *France* and this was manifest by the total, almost funereal silence of the vast crowds who were there to witness her leave-taking. Equally, not one ship in the port sounded its siren in farewell in response to *France* when she forlornly sounded her horn three last times. The ship had no motive power of her own since only a small number of her boilers had been reactivated to provide electrical services such as for her winch gear, steering, on-board lighting and radio communication with her escort. After a slow passage the convoy, for there were now more tugs around her, arrived at Bremerhaven on 22 August.

It was to be a 'root and branch' facelift and refurbishment for *Norway*, at a cost of $80 million. The renowned Danish naval architect Tage Wandborg was engaged to design and oversee the project, which was in all respects a major challenge. As one observer put it, the task was to change a two-class 'indoor' liner, designed for the rigours of the North Atlantic, into a single-class 'outdoor' recreational cruise ship for operation in the more tranquil waters of the Caribbean. Given the scale of the modifications, the transformation was completed in an extraordinarily short period of time, just thirty-three weeks in the shipyard and six weeks ahead of schedule. This is in stark contrast to the three years that the renovation of *Liberté* had taken, even allowing for the flooding damage that the earlier ship had sustained. Kloster's philosophy in purchasing *France* for conversion was to offer the greatest possible enjoyment to the greatest number of people on the greatest ship in the world.

Structurally, the modifications appeared, on the face of it, to be superficial but were in fact significant and extensive. The upper deck and promenade deck were extended aft-wards, eliminating the two small masts on either side of the promenade deck. The promenade deck aft was also widened to provide as much space as possible for outside, leisure-orientated facilities, extending almost to the stern and projecting out over the ship's sides. The sun deck was also broadened, taking it over the lifeboat davits on the boat deck. As a result, these modifications increased the outside spaces by 70,000 sq ft. Exploiting the greater area, a new heated outside pool was installed on the promenade deck extension, now named the pool deck.

At the forward end, the turtle back and breakwater on her foredeck were eliminated, as were the two derricks. In their place were two immense davits supporting a pair of large 55-ton passenger tenders, *Little Norway 1* and *Little Norway 2*, measuring 93ft in length and each capable of carrying 400 passengers. Their purpose was to ferry passengers between ship and shore on port and island visits.

To reduce operating costs, a conversion to diesel engines was contemplated but as this would have been prohibitively expensive it was determined instead to retain her steam plant, or at least part of it, with the remainder decommissioned and left undisturbed within the hull. As an operating speed of 16 to 18 knots was considered to be adequate for her cruise ship role, power output of 40,000 shp was sufficient with a reserve maximum of 54,000 shp for a top speed of 25 knots. Thus, only four of her eight boilers were retained in an active state that with the associated turbine machinery were to be fully automated, managed from a central control point.

A contentious suggestion was to remove the ship's distinctive funnels since only one was required with four of the boilers decommissioned, and replace them with a single new funnel aft. Mindful of how her funnels were truly emblematic and made the former *France* instantly recognisable, this unsavoury proposal was rejected swiftly. However, in the light of later events, the retention of the steam plant may have been a short-sighted mistake, even though fuel consumption with a diesel propulsion arrangement would have been higher. Running on reduced boilers, *Norway*'s daily fuel consumption was 250 tons, some 66 per cent less than it had been for *France*.

Her two outside propellers were dispensed with, their drive shafts plugged and the shaft brackets removed. Although inactive, the forward turbine machinery was left in situ in order to retain stability but part of the redundant forward boiler room was cleared to accommodate a diesel-powered electricity generating system. For manoeuvrability, and to avoid dependence on tugs for port handling, five sets of bow and stern thrusters were fitted, three forward and two aft.

Internally, considerable changes were also made, although some of the former public spaces were retained, given a makeover and renamed. Others were completely restyled to reflect the new cruising role of the ship. A significant alteration, in order to adapt *Norway* into a one-class vessel, was the upgrading of all the former tourist-class cabins to a first-class standard, rather than the downgrading of the original first-class accommodation. Either side of the former promenade deck, now rechristened the international deck, remained closed in with cafés, bistros and boutique shops along their length, the starboard side named the Champs Elysées, the port side given the name Fifth Avenue. Above, on the Olympic deck there was a circuitous open promenade.

The existing restaurants remained largely unchanged but to further enhance the al fresco ambience that was intended for her cruise passengers, a self-service open-air restaurant was constructed across the full width of *Norway*'s international deck. All the cabins were air conditioned, had a television set, a telephone and a bathroom with shower and hand basin, but not all had either a porthole or window, a shortcoming when compared to modern, purpose-built cruise ships.

The changes made to *Norway* raised her tonnage to 70,219 gross, allowing her to regain top spot as the world's largest passenger ship by overtaking the revised figure of *Queen Elizabeth 2*. Her overall length remained unaltered but her maximum beam increased by 4in! The revised passenger capacity of *Norway* was 1,944 and her crew, reduced from her days as *France*, was 907, of whom 107 were Norwegian, both officers and deck grades.

On the completion of the work, *Norway* sailed to Oslo, where she was formally christened on 3 May 1980 in the presence of King Olav V, Prince Harald and Princess Sonia. Oslo became her new port of registration until 1987, when it was switched to Nassau in the Bahamas. Following the renaming ceremony, *Norway* was opened to the public and was visited by some 40,000 people. She then set out across the Atlantic for repositioning, calling at Southampton and New York en route and arriving at Miami, her new base port, on 22 May 1980.

Ready to commence her new role, somewhat different to other cruise ships on the Caribbean circuit, *Norway* typically operated seven-day cruises out of Miami that were characterised by dependence on her on-board facilities, treating the ship as a destination in itself. Typically, she made calls at Key West, Cozumel in Mexico and Roatán in Honduras or Saint Thomas and Nassau, with a stopover at a private island owned by the Norwegian Cruise Line, either Little San Salvador or later, Great Stirrup Cay.

Subsequently, *Norway* undertook a programme of European cruises, conveniently combined with her regular overhauls at Bremerhaven. In April 1982, a more involved technical transformation was initiated that lasted three months, during which the redundant boilers in the forward engine room were dismantled and additional diesel generators installed in their place. Other changes were replacement four-bladed propellers and the incorporation of satellite navigation capability. Again in 1984 and 1987, *Norway* returned to Bremerhaven for attention, more modifications and the upgrade of some of her staterooms.

As time went on, larger purpose-built cruise ships entered service, the momentum of this upward trend having been vindicated by *Norway*'s success. These new vessels, equipped with a greater range of amenities for cruise passengers, represented a challenge to the now ageing *Norway*. The commissioning of Royal Caribbean Cruise Lines' 74,000 gross ton *Sovereign of the Seas*, *Majesty of the Seas* and *Monarch of the Seas* from 1987 relegated *Norway* to fourth place in the size order. Carnival Cruise Line, another fast expanding cruise operator, brought out its own comparably sized ships and it was clear that *Norway* was fast being left behind. Confronting the competition, Kloster had *Norway* returned to Bremerhaven where, between 3 September and 2 October 1990, she had two new decks built atop the existing superstructure, increasing her passenger capacity to 2,565 and her gross tonnage to 76,049. Once more her position as the world's largest passenger ship was restored. But it was short-lived. Princess Cruises introduced the bigger *Sun Princess* and her sisters from 1995 and when *Carnival Destiny* emerged in November 1996 at over 100,000 gross tons, *Norway* was left well behind by that measurement.

A refit in the King George V drydock at Southampton in 1996 to upgrade her to the latest SOLAS standard coincided with a change to her funnel colours. The wavy design of white and two shades of

The former *France* under conversion into *Norway*, at the Lloyd Werft shipyard on 8 January 1980. (Peter Nath)

As first converted and in her original Norwegian Cruise Line colours, *Norway* makes her inaugural call at New York. (Bill Miller collection)

blue was substituted by a navy blue livery upon which the letters NCL within an upright rectangular outline were painted in gold. More than that, the function of her remaining active funnel was also altered. The boiler fumes no longer exited via the funnel wings but through pipes positioned in the funnel base, leaving the wings as no more than a decorative feature.

Like *Normandie* back in June 1936, *Norway* was also the centre of an accident involving aircraft flying too close to her. On 30 July 1998, while she was anchored in Quiberon Bay, Brittany, two aircraft, a Cessna 177 and a Beechcraft 1900D, circled low over *Norway*. In the risky manoeuvre that followed, the Cessna struck the other plane, causing both to crash into the sea and killing all fifteen people on the two aircraft: the fourteen passengers and crew aboard the Beechcraft and the lone pilot of the Cessna.

As the place of *Norway* in her owners' fleet declined, so too the extent of her maintenance and upkeep was also reduced. She became a far cry from the almost contemporary *Queen Elizabeth 2* that, afforded an appropriate level of care and attention, was operating five-star cruises. In contrast, as her mechanical and structural problems mounted, *Norway* was relegated to bargain-basement cruises, albeit still well prescribed. To make matters worse, *Norway* began to experience an increasing number of problems: a turbocharger fire at Barcelona in 1999, propeller shaft leaks off Norway and then a sequence of boiler breakdowns. She had previously experienced an extreme boiler failure, back on 1 May 1981, which was perhaps a portent of future mechanical issues. It required the complete shut down of her main engines, leaving the ship at sea powerless with a complement of 1,787 passengers on board. Two other NCL cruise ships, *Starward* and *Sunward II*, were instructed to proceed to Nassau and Freeport to disembark their passengers and then make their way to assist the crippled *Norway* by taking off her passengers and transferring them to Miami. After two days, with two of her boilers restarted, *Norway* limped into port after her fleet mates. Her following cruise was cancelled and she did not resume service until seven days later.

The litany of shipboard incidents increased, to which were added matters of safety violation, imposing port detentions, and illegal dumping, occasioning fines on her owners. In 2001, *Norway* failed a coastguard inspection, was deemed to be hazardous in the event of an on-board fire and was therefore detained in port until repairs and mandated corrective actions were complete.

These woes culminated in a major, fatal boiler explosion shortly after *Norway*, inbound from a seven-day cruise, docked alongside piers 1 and 2 on Dodge Island, Miami, early in the morning of 25 May 2003. There were 2,135 passengers and 911 crew aboard. Eight members of the crew were killed and ten severely injured, while a further seven sustained less serious injuries.

The explosion happened during the shut down of her engines. Boiler 23 ruptured, instantly releasing superheated steam initially into the engine room spaces. Four of the engine watch crew who had come on duty at 0400 hours were killed almost instantly. The four D-type water-tube boilers were located aft on deck 2 in the next-to-last vertical compartment, extending upwards through deck 3 to the Caribbean deck, deck 4, because of their height, the boiler casing ending at deck 5, the Biscayne deck. The boiler room and casing were surrounded by bulkheads intended to withstand the spread of fire and smoke for a minimum of sixty minutes. Each boiler contained 20 tons of water at 528°F and 870 psi. At the time of the incident, boilers 22 and 24 on the port side and 23 on the starboard side were operational, while boiler 21 was not in use. With the violent explosion accompanying the rupture of boiler 23, the pressurised water expanded within seconds into superheated steam at normal atmospheric pressure, its volume increased by a factor calculated to have been times 1,260. Forced upwards carrying smoke, soot and debris, it breached the bulkheads, allowing the steam to penetrate into the adjacent crew quarters on decks 4 and 5, extensively scalding six off-duty crew members located there. Seven other crew members, scalded over 20 per cent of their skin surface, managed to escape by jumping overboard with the intention of reaching the pier, 8ft below. Of these, three plunged into the sea and the remainder landed on

This stern view of *Norway* shows the specially incorporated deck extensions aft that projected out beyond her hull sides. (Jean M. Otten)

the pier although in doing so one suffered a broken leg. Over the next fifteen minutes, another 125 crew members who were largely unscathed, left the ship via a gangway on Biscayne Deck.

The nearest passengers were occupying cabins on Biscayne deck, relatively close to the bulkhead breach, but immensely fortunate there were no passenger casualties. The loss of all the boilers when the boiler automation system shut them down led to the interruption of steam supply to the turbogenerators, knocking out all electrical power. Fortunately, however, within minutes the emergency diesel generators were activated. Fifteen minutes after the explosion, having despatched on-board fire and medical teams to the scene and alerted the City of Miami Fire Department, *Norway*'s master, Captain Haakon Gangdal, addressed the passengers over the PA system, directing them to immediately make their way to the lifeboat muster stations.

While shoreside firefighters were boarding the ship, a passenger count was made that initially revealed that forty-one passengers were missing. The PA announcement to proceed to muster stations was repeated and by 0800 all passengers were accounted for, permitting disembarkation via a gangway on Norway deck and by 0900 all were ashore. At that point, the master ordered the sounding of the abandon ship alarm and the remainder of *Norway*'s crew vacated the ship. Apart from the known casualties, a roll call at 1045 was able to confirm that there was no one left aboard. The emergency having been stabilised, the crippled *Norway* was left secure and empty, neither able to resume operation nor permitted to do so.

The US National Transportation Safety Board subsequently carried out a marine accident investigation. Its report, released over four years later, on 29 October 2007, was highly critical of Norwegian Caribbean Line, *Norway*'s engineers and Bureau Veritas (BV), the survey and registration authority. The boiler failure was identified as having occurred at the waterfall header at the base of the combustion chamber. Questioned in the inquiry, the Second Engineer, the most senior officer on duty at the time, who had been monitoring the boiler gauges and instruments in the engine control room aft of the boiler room, later stated, 'All was normal.' No alarms were indicated, he said, and he had heard only 'normal engine/boiler room sounds'.

The NTSB report summary declared as follows:

> The Safety Board determined that the following factors contributed to the rupture of boiler No. 23 on the *Norway*:
>
> - Lack of adherence to water chemistry composition limits and procedures by both the water chemistry subcontractors and NCL during wet lay-up periods, leading to pitting from oxygen corrosion.
> - Failure to take number of boiler cycles into account during maintenance.
> - Severe thermal transients from heating and cooling the boilers too quickly and from constraints created by frozen boiler support feet.
> - Use of questionable weld repair procedures.
> - Lack of appropriate non-destructive testing by the BV surveyors and NCL inspectors to determine whether cracks were present.
> - Inadequate survey guidance from BV to its surveyors.
> - Failure to repair cracks into which copper had been inappropriately introduced.

The idle *Norway* languished at Miami for almost four months until she was towed away to Bremerhaven, where she docked on 23 September 2003. For a time, deliberations ensued as to whether she should be either repaired or permanently taken out of service until, finally, NCL's Chief Executive, Colin Veitch, announced on 23 March 2004 that *Norway* would 'sail no more' and would be disposed of for breaking up.

However, the saga did not end there for, with large quantities of asbestos within her, she was designated as a source of hazardous material and under the Basel Convention the German government was not permitted to release *Norway* to be taken away without binding assurances that she would be dealt with in accordance with the stipulations of the convention. Another year passed.

Subsequently, the German authorities granted exit permission on the firm understanding that *Norway* was not to be broken up but refitted in Asia for service in Australia. Still under the Star Cruises (Genting Hong Kong) banner, the parent company of Norwegian Caribbean Line, by then revamped as Norwegian Cruise Lines, the ship left Bremerhaven on 23 May 2005 under tow of the ocean-going tug *De Da* and arrived off Port Klang, Malaysia, on 10 August. Breaking up was clearly the intention and it was argued that the assurances made to the German authorities had been fraudulent and duplicitous.

Norway after her final modification and in her third new colour scheme. (Fabien Trehet)

Two unrelated campaign groups now became involved in *Norway*'s affairs, both pursuing efforts to prevent her demolition but for different reasons. One group was seeking her preservation, suitably sanitised, at a location in France. The other, led by Greenpeace, was fighting to have breaking up of *Norway* prohibited under international law, as long as the hazardous substances aboard her had not been removed, as legally required, and the ship verifiably decontaminated.

In April 2006, the hapless *Norway* was sold to Monrovian interests, who renamed her *Blue Lady*. The name *Norway* on her hull was obliterated by paint and the new name crudely stencilled over it. Around this time, the government of Bangladesh, one of the potential ship-breaking destinations for *Blue Lady*, refused entry into their waters because of the legal and environmental issues. The ship, still a centre of uncertainty, was shuttled from Malaysia to Fujairah, in the United Arab Emirates, and then to Alang, India. Despite international protestations, and after a dubious inspection, the Indian government issued entry clearance for *Blue Lady*. However, further litigation followed amid a campaign to have Star Cruises, and its subsidiary Norwegian Cruise Lines, prosecuted for corporate negligence. In the end it did nothing to prevent the inevitable. On 11 September 2007, it was declared that the former *France*, former *Norway*, was safe to scrap and she was beached at Alang. The debacle and the deception surrounding the questionable final decision did little for the credibility of India.

On 4 December that year, a section of the ship's bow was removed as a symbolic indicator that her life was at an end. Full-scale scrapping commenced from January 2008 and by the following December the demolition had been essentially completed. The tip of her bow, dubbed *le nez*, was purchased and returned to France, where it is displayed alongside the River Seine by the Paris Yacht Marina at the Port de Grenelle.

The prolonged fiasco attending her final years amounted to a pitiful degradation, clouding the career of this once magnificent super-liner. To date, of all the great transatlantic super-liners, she accumulated the greatest number of active service years, at a total of forty-one while operating under the names *France* and *Norway*.

EIGHT

A SHIP FOR ALL OCEANS AND A NATIONAL TREASURE: *QUEEN ELIZABETH 2*

Despite a discouraging operating turnover and profit in 1957, the Cunard Line, under the chairmanship of Colonel Denis Bates, was giving serious consideration to replacing *Queen Mary* and *Queen Elizabeth* when they came to the end of their operational careers. They still had more than a few years left in service but the question of replacement had to be addressed and to this end an independent committee of three, the Chandos Committee, was appointed by the British government. They were to advise on replacement plans and to consider the £40 million that was requested by Cunard for this purpose. Proposals were discussed for two ships of around 75,000 tons and 1,040ft in length, with a mean speed of 30 knots, costing £20 million each.

The project ground on and the government felt it was important that the UK should be competitive on the North Atlantic; after all, in the United States the North Atlantic liner services were subsidised by 58 per cent and in France by 20 per cent. However, the Chandos Committee proposed only one ship to replace *Queen Mary* and for its construction the Treasury would loan up to £18 million at 4½ per cent over twenty-five years rather than subsidise the construction. The new vessel was to carry 2,270 passengers with a speed in the range of 30 knots, at a total cost of between £25 million and £28 million, but not exceeding £30 million. Cunard would provide equity capital of £12 million. At this stage the two *Queens* had been written off, with their scrap value at around £800,000 each.

The replacement proposal seemed plausible enough but in reality the British taxpayer would be required to stump up the £18 million whether they liked it or not. The loan was to be interest free during the anticipated construction time of four years, with a special proviso that any government money drawn during the first two years would bear interest at 2½ per cent up to the time of the replacement liner's maiden voyage. Thereafter the whole loan would bear interest at 4½ per cent. The replacement *Queen* was designated *Q3*.

As things panned out, the technical specifications from the Cunard design team became more apparent. The proposed gross tonnage was required to be around 75,000 (or even up to 80,000). *Q3*'s length varied from 1,040ft to 990ft, with a beam of 116ft and a draught of 35ft. Her propulsion was to be four sets of steam turbines driving four screws producing 200,000 shp, with steam supplied by eight boilers at 850 psi and 950°F. *Q3* would essentially be a three-class North Atlantic ferry designed specifically for that route.

The technical dimensions arrived at were much debated. Aspects considered included that, for a given speed, a certain minimum waterline length was required in addition to the force of the engine power. Therefore, for an average speed of 33 knots, a length of 900ft would suffice. If the replacement liner was scaled down along these lines she would work out at around 50,000 gross tons instead of 75,000, about the size of *United States*, which had been designed and built as a fast US Navy transport. *United States* had a known speed of around 40 knots, which required more than twice the shaft horsepower needed for a 30-knot liner while employed on the North Atlantic run. In effect she had been forced to carry a surplus of heavy and costly machinery, with its US Navy 'redundancy' requirement, to the detriment of her passenger and freight capacity. A vessel of the *United States*' class, where naval employment took precedence over peacetime working, was not the most profitable type of commercial investment.

Queen Elizabeth 2 on the slipway at Clydebank, showing her flared bow extending down to its bulbous forefoot. (Cunard)

A change of direction was perhaps on the cards since Sir John Brocklebank, who took over the chairmanship of Cunard early in 1960, favoured a dual-purpose passenger ship. During March 1961, tenders for *Q3* were invited from prospective UK shipbuilders, even though, ironically, between January and March of that year only 91,300 people had crossed the Atlantic by ship. The following 26 May *Queen Mary* was 25 years old and the need for a decision on her replacement was becoming pressing.

Between 28 and 31 July 1961, five bids had been submitted from the six leading shipyards. However, that same month Raymond Gregory, a Cunard shareholder who was unhappy about the traditional Atlantic ferry role of *Q3*, canvassed the 20,000 Cunard shareholders to reject the proposed design as being out of date, in favour of a dual-role ship. It transpired that 4,227 were in support of Gregory's proposals, while 297 subscribed to the *Q3* design. Gregory found an ally in Sir John Brocklebank and on 19 October 1961, following a shareholders' revolt, the order for *Q3* was postponed indefinitely and later cancelled.

History will judge this decision as being the right one. Now that the battle for *Q3* was over, the battle for *Q4* was about to begin. It was a classic case of 'back to the drawing board'. From the *Q3* investigations it was evident that marine technology had advanced to a situation whereby the required passenger complement and speed could be achieved in a smaller ship driven by twin screws.

It was not just a simple case of recycling a design based on the footprint of the *Q3* plan that had been consigned to history but the need for a radical new departure, a design for the 1970s and '80s. One that would reflect the 'swinging 60s' as well as continue to impress well into the future in a similar vein to that which had 'wowed' the travelling public when John West's design for *Canberra* entered service in 1961 and French Line's *France* when she entered service in 1962. They belonged to the space age and seemed straight out of Dan Dare's comic strip in the boys' magazine *Eagle*. Tastes, trends and styles had to be taken into consideration besides a dual-purpose super-liner having to pay for itself way into the future. To this end Cunard's design team under Senior Naval Architect Dan Wallace came up with the *Q4* design for a steam turbine powered twin-screw liner, not intended to be a Blue Riband contender. The team comprised Tom Kameen, the Technical Director, W. Marcus, F. Wolstenholme and R. Hutchinson. The ship's external profile was left to James Gardner, who had previous experience with the Royal Festival Hall architecture and design.

In January 1963 a further round of negotiations was brokered with the government about a loan, which was rejected and Cunard was referred instead to the Shipbuilding Credit Scheme. At this meeting some shareholders were starting to get cold feet, when Raymond Gregory came out with the bold statement: 'Either Cunard must build this ship or it must die.'

In light of this, Cunard's revised requirements called for such a liner of reduced size, suitable for off-peak cruising in both the Atlantic and Pacific, able to transit the Panama Canal and with construction costs nearer to £22 million. Already described as *Q4*, this specification ultimately resulted in *Queen Elizabeth 2*. At a shareholders' meeting in July 1963, Sir John Brocklebank promised a decision on *Q4* in 1964. After weeks of delicate negotiations, on 21 October 1963 Sir John announced that an agreement had been signed whereby £17.6 million could be made available at 4½ per cent from the Shipbuilding Credit Scheme. Apart from this, after much discussion the new ship would be designed as a two-class ship, her features to include stabilisers, air conditioning throughout and a lido deck.

On 30 December 1964 the order for the ship was placed, with *Queen Elizabeth 2* eventually entering service during April 1969. Hereafter the ship was referred to affectionately as *QE2*. The evident advantages of her design that had been advocated by Sir John were a gross tonnage measurement of 66,863, an overall length of 963ft and a draught of 32ft 6in. This compared favourably with *Canberra* and *France*, both of which had a draught of 34ft 6in. These two liners were somewhat restricted on the cruise circuit as a result of their draughts, which precluded them from some essential

Launch of *QE2* on 20 September 1967. Unlike *Queen Mary*, her name was not kept secret until launch day. (Cunard)

Nearing completion, *QE2* lacks her lifeboats and the final coat of paint on her superstructure. Her controversial funnel design had been revealed by this point. (Cunard)

berthing facilities. *QE2*'s beam, at 105ft, was some 15ft less than the aborted *Q3*, enabling her to transit the Panama Canal. In addition to this she had a high-capacity distilling plant that enabled her to make voyages on longer routes than the North Atlantic. The old *Queens*' fresh water tanks capacity limited their time at sea to seven days!

Q4 would be steam turbine driven in the traditional mode attracting the very latest Pametrada turbine design configuration. Initially *Q4* was to have four boilers and five generators located in two different compartments in the interests of safety. However, to reduce overall costs a revised machinery layout was agreed upon, employing three boilers and three generators. With the benefit of hindsight the choice of three boilers meant that there would be no 'slack in the system' (also known as 'redundancy') in case of standby, breakdown or failures.

The boilers, Foster-Wheeler External Superheat Drum II type, the largest installed in a passenger ship, were chosen for their more advanced steaming conditions of 850 psi at 1,000°F, other than for a two to four-hour trial period when the ship operated at 950°F.

The main engines were manufactured by John Brown & Co. to the Pametrada design, consisting of a pair of double-reduction, cross-compound turbine sets each capable of producing 55,000 shp maximum. Although the designed service speed only required a total of between 85,000–95,000 shp, Cunard considered it better to have the maximum power that could be prudently absorbed by twin screws.

Cunard carried out a very comprehensive exercise to assess the marketing resources encompassing the ship. In particular, the state of readiness and proposed interior design of the passenger terminal were scrutinised. During September 1963 the specifications for hull, machinery and electrical equipment were put out to five major shipbuilders with invitations to tender by the end of November. John Brown of Clydebank, a Swan Hunter & Wigham Richardson/Vickers-Armstrongs consortium and Harland & Wolff all tendered. A contract price of £25.43 million was submitted by John Brown and Sir John Brocklebank, and Lord Aberconway signed the contract with the yard on 30 December 1964, with a delivery date of May 1968.

John Brown had invested some £5.5 million since the end of the Second World War to modernise the shipyard, introducing new progress planning to bring it up to date. At the building berths, where much renovation had taken place, vastly improved cranes were installed. Also renovated were the prefabrication sheds, where assembled units of up to 110 tons could be built before being craned into position. In addition, by the use of computers, steel plate could be flame cut by the acetylene cutter following the architects' drawings by a 'magic eye' system. John Brown's craftsmen were fully trained to handle these new methods and materials.

The superstructure was all-welded aluminium alloy, based on experience gained with *United States*, *France* and *Canberra*. The hull was of all-welded steel, where the plates were welded edge to edge, thus replacing overlapping plates, riveting and caulking. The exception was, as with the previous mentioned liners, where the aluminium joined the steel such that the two metals had to be insulated.

The power plant selection was to be steam turbines of the Pametrada design driving twin propellers to maintain an average speed of 28.5 knots on the North Atlantic. The six-bladed screws, fabricated of manganese bronze, were designed to absorb the maximum of 55,000 shp at 174 rev/min, the number of blades settled on after the vibration results of the ship and hull had been analysed.

The HP turbines comprised of one Curtis Wheel, followed by eleven impulse stages, had a maximum rotational speed of 5,207 rev/min, whilst the double-flow, impulse-reaction LP turbine rotated at 3,192 rev/min. For astern manoeuvring the company insisted on a complicated arrangement of two turbines connected in series, one on the HP pinion and the other on the LP. A single HP astern Curtis Wheel was placed on the forward end of the HP shaft whilst two sets of astern blades were placed in the LP casing. Full astern power could be obtained from a single Curtis Wheel and a single row of blades.

Turbines and boilers occupied separate rooms, the boilers being forward of the turbines. In this manner two of the boilers could be installed abreast the beam of the ship and one abaft of these two. Had four boilers been installed it would have involved extra costs because the combined lengths of the engine room and boiler rooms would have exceeded the lengths that could be allowed in side damage calculations. The vacant space resulting was occupied by an ancillary plant. The choice of steam turbines was interesting in that they may have been at the zenith of marine propulsion of this type by this time. The previous passenger liner completed by John Brown in 1966, the Swedish America liner *Kungsholm*, had been installed with twin 9-cylinder Gotaverken diesel engines. (Sadly, John Brown's had lost £3 million on the building of the *Kungsholm*, on a fixed-price contract, which was four months late at delivery).

Ironically, Cammell Laird at Birkenhead had manufactured their last steam turbine set in 1965 for the BP Tanker Co.'s *British Captain*.

Considering the precision in the construction of steam turbines, with their many stages of turbine blades, both static and rotating, coupled to meshing gears, it required all these components to function correctly. But if just one component malfunctioned it might spell disaster.

Aerial view of *QE2* while fitting out at John Brown's, Clydebank. (Cunard)

The construction of *Q4* at Clydebank took place against a background of industrial decline, as well as economic and social turmoil. Back in the mid-1960s shipbuilding was very much a bespoke industry with a plethora of trades, restricted working practices, poor health and safety practices, pilfering, and trade union demarcation by the construction and outfitting trades, along with uncovered building berths that were open to the weather. A large, mainly male workforce was overseen by bowler-hatted foremen, shop foremen and yard managers. It was a melting pot of industrial humanity into which the construction of *Q4* was plunged, amid skilled, semi-skilled craftsmen and unskilled support, where some 3,000 men worked on the construction with a further 500 at the fitting-out period. What was not seen or fully appreciated was a covert undercurrent of a 'them and us' scenario.

Following in the footsteps of the previous *Queens*, *QE2* sails along the River Clyde to the open sea. (Cunard)

Rejection! A forlorn *QE2* alongside the Ocean Terminal, Southampton, during January 1969. Her turbines were undergoing repair and further remedial work was being carried out by Upper Clyde Shipbuilders. Aft of her, to the right, is the former Irish ferry *Cammell Laird*, ex *Royal Ulsterman,* on which the builder's workforce was billeted. (Richard de Kerbrech)

The first keel plates of Yard No. 736 in the form of a 180-ton prefabricated module, 117ft long by 23ft wide was laid on 5 July 1965, on the same slipway where the two previous *Queens* had been built. At this stage it was hoped that the *Q4* would be in service by spring of 1968. Four months later, on 8 November 1965, Sir John Brocklebank resigned as Cunard's chairman and was replaced by Sir Basil Smallpeice.

During the construction, 1,100 tons of Alcan aluminium alloy were delivered to the yard. To prevent any corrosion between the steel hull and the aluminium alloy superstructure, a careful bonding method had to be used. A special epoxy resin compound was spread along the joining surfaces between the two metals and then steel rivets were used as the fastening.

Dan Wallace and his Cunard design team worked well in concert with John Brown's design team headed by J.F. Starks, Chief Naval Architect and the company's Assistant Managing Director. As work progressed, however, the financial position of John Brown's shipyard continued to deteriorate. Yard No. 736 had been laid down before the true scale of losses on the Swedish America's *Kungsholm* had been revealed.

Although the work made good headway, John Brown advised Cunard that *Q4*'s launch was scheduled for September 1967 and delivery in November 1968, some six months late in both cases. From the initial laying of the keel, the erection of steelwork on No. 4 slip had proceeded at a steady pace. Extra men were taken on when necessary who, if they did not have any documentary evidence to confirm their competence at their trade, were given a skills test.

To maintain an adequate supply of steel such as plate, angle bar and tee bar, the steel mills in the vicinity of Clydebank would give advance notice of their rolling dates for particular forms of supply to the drawing office, so that sufficient time could be allowed for ordering. Because *Q4* was practically all welded, the traditional sounds of riveting and caulking no longer sent their staccato rhythm over Clydebank.

On the morning of 20 September 1967, *Q4*'s newly painted hull loomed large over her surroundings, an imposing and massive vessel with the underwater body of a large whale-like shape, fine of form, with a sheer and flair, all set for her working environment no matter what the weather. In the past only those who had made the wax test tank models were able to appreciate the ship's underwater hull form.

QE2 as she appeared when she first entered service. (Cunard)

At the fore end was the ship's bulbous bow, the first ever built on a Cunard ship, its purpose to push a pressure wave ahead of the ship cutting down the wave-making resistance on the hull and reducing fuel consumption. Moving along the full form of the ship, viewed from the Clyde, were the twin, six-bladed manganese-bronze propellers poised above the high tide. Each was 19ft in diameter and weighed 32 tons. Unlike on *France* with her exposed propeller shaft 'A' brackets, *Q4* had enclosed shafts.

With thousands gathered at the yard and along the banks of the River Clyde, and TV and Pathé Newsreel crews present and ready to roll, at 2.30 p.m. Her Majesty Queen Elizabeth II stepped forward and christened *Q4*, '*Queen Elizabeth the Second*'. As the mighty hulk slid down the ways into the Clyde there was a united sense of pride and 'them & us' fleetingly became one. This was a ship of national importance about to embark on a forty-plus-year career.

In the meantime, Cunard's *Mauretania* had been withdrawn from service in 1964 and from 16 May to 1 July 1966 the National Union of Seamen held a national strike, paralysing British merchant shipping and affecting sailing schedules. The following year, during October 1967, *Queen Mary* was sold to Long Beach, California. The Cunard fleet was in a state of flux as *QE2*'s construction advanced.

It is a peculiar trait of the British to take an individual or product of great achievement and place it on a pedestal and laud it. But then, as soon as any bad publicity comes along it is denigrated, no longer praised or admired. Such was the case pertaining to some of the events surrounding *Queen Elizabeth 2* during her long service career.

Upon her completion there was much talk about the radical funnel design that departed from the traditional. James Gardner was responsible for styling the external appearance of the liner and, after many deliberations regarding the design, he came up with the scoop style, which was successfully wind tunnel tested. Also, besides its unusual profile, it was not painted in the traditional Cunard funnel livery but in a grey, black and white theme.

On 26 November 1968 the new super-liner underwent preliminary speed trials off the Isle of Arran, during which she achieved 29 knots.

It was almost farewell to the Clyde when, on 30 November, the trials were terminated because oil fuel had contaminated the steam and feed water system. She was drydocked at Greenock for the system to be cleared, delaying further acceptance trials. In the meantime, the John Brown shipyard had been absorbed into the Upper Clyde Shipbuilders (UCS).

When the clearance of the feed system had been completed, *QE2* sailed for the Canary Isles for further proving trials but, on 24 December 1968 while on these trials, the starboard HP turbine started to vibrate and later a grinding noise was heard from within. On Boxing Day, the port HP turbine was experiencing similar symptoms. Repairs could not be affected at sea, so with a loss of power *QE2* limped back to Southampton, arriving there on 2 January 1969 after a three-day journey.

As a consequence, Cunard refused to accept her and no one knew the extent of the damage. By early January the lagging was removed, the turbine covers were lifted and the rotors inspected. Hundreds of blades had been stripped from the rotor hubs. A complete row of blades from No. 9 stage and several from stage 10 had been broken off and lay in the bottom of the casing. Further inspection revealed cracking at the roots of the blades. In the port turbine the damage was less severe; seven blades from stage 9 had broken off and there was some cracking in stages 8 to 10. The rotors were removed and sent back to John Brown Engineering.

By this time many craftsmen in the finishing and outfit trades had been flown down from UCS and were billeted on a former Irish ferry, *Cammell Laird*, moored aft of *QE2* in the Ocean Dock. Much work was carried out by these men, augmented by local tradesmen from Vosper-Thornycroft. In all some 1,400 men worked on the ship. The consultant engineers' findings were that the blade roots, which were square, did not expand as much as could be expected and that the steam supply nozzles, used to jet steam to the blades, were below standard; also there were more than needed. The resultant vibration and resonance induced in the blades proved to be too much and as one cracked and broke off it took away

Alongside at Southampton's Ocean Terminal, *QE2* seen from above. (Cunard)

adjacent blades until many of them had been torn from the rotor. (In a similar manner to the expression 'the shit hitting the fan'.) The remedy was to provide new blades with strengthened roots and to 'tie' them together with a circumferential length of ⅛in stainless steel rod, brazed between the blades of all stages, in order to supply stiffness and a resistance to resonance.

The entire incident caused massive reputational damage to John Brown & Co. at a critical point in Clyde shipbuilding, at the very time when they had been merged into UCS. Despite these initial 'teething' problems, however, these turbines continued to propel *QE2* for 2.6 million nautical miles over eighteen years, or 144,444 miles annually! Sadly, Pametrada, the turbine designers, had gone out of business under the recommendations of the government-commissioned

Geddes Report, which had also brought about the UCS merger. It is of interest to note that according to Ingvar Jung's monograph *The Marine Turbine*, he states on the blade failures: 'This classic error, which General Electric of the USA suffered painfully from during the twenties and de Stal Laval in the thirties, and which put paid to the Brown-Curtis turbines in Britain during the twenties and thirties had apparently been forgotten by Pametrada.'

After many weeks of wrangling between owners and builders, by the end of March 1969 the refurbished turbine rotors were reinstalled and the shipbuilders assured Cunard that she would be completed two weeks ahead of the revised estimated date of 2 May. Further tethered turbine trials and drydocking then took place, followed by a small shake-down cruise to the West African coast. On 18 April *QE2* was finally handed to Cunard; final cost £29.09 million.

QE2 departed Southampton for Las Palmas on 22 April on her first voyage with fare-paying passengers and then on 2 May she sailed on her maiden departure for New York with 1,400 on board.

To relate each and every voyage of *QE2* and chronicle every incident that befell the liner is beyond the scope of this publication, so only certain salient incidents will be highlighted hereafter.

QE2 outbound in The Solent after rounding the Bramble Bank. (Peter Bridle)

On the evening of 8 January 1971 around 5 p.m., the French Line's *Antilles* went aground on an uncharted submerged reef half a mile off Mustique, an island co-owned by brewery chief Colin Tennant. Her oil tanks ruptured and oil spilled onto the surrounding sea. It also entered the boiler room, where it ignited. At one time the firefighting crew seemed to have the outbreak under control but the flames spread with such intensity that the fire had irrepressibly taken hold. At 5.22 p.m. the *Antilles*' distress call for assistance was picked up in Barbados. The 635 crew and passengers aboard her abandoned ship and all were landed safely on Mustique. Upon answering the distress call *QE2* reached Mustique on 9 January and embarked over 500 of the survivors and landed them at Barbados for onward destinations. The remainder reached the island by the ship's lifeboats or aboard the French banana boat *Point Allegre*.

Since entering service, operation of *QE2* had barely broken even, indeed she had started losing £500,000 a year. Cunard was acquired by Trafalgar House in 1971 and Sir Basil Smallpeice and sixteen other senior Cunard executives were paid off with golden handshakes as Trafalgar's Victor Matthews and his team took over.

On 17 May 1972 a telephone call was made to Cunard's New York office in which the anonymous caller stated that unless a ransom of $350,000 was paid in $20 and $10 notes, two accomplices aboard the ship, one terminally ill with cancer, would detonate six explosive devices and depth charges hidden within the liner. The FBI was called and Cunard alerted the Ministry of Defence and Scotland Yard in London. The money was prepared with which to pay the ransom in New York, while in the UK the RAF scrambled a Hercules transport aircraft to fly out a bomb disposal team accompanied by a Nimrod reconnaissance plane to handle communications. The four-man bomb disposal team, comprised of specialists from the Army, SAS and the Royal Marines Special Boat Service, who rendezvoused with *QE2* while she was en route to Cherbourg. Captain William Law addressed the 1,438 passengers on board over the PA system, advising them of the situation. *QE2* hove to as the Hercules parachuted the team into the sea to be picked by a ship's motor boat. All this provided a great spectacle for the passengers, who lined the rails. The military team then carried out a thorough search of the passenger accommodation, engine room and baggage room. Fortunately, the bomb scare turned out to be a hoax but it was all credit to Cunard, who had put the safety of the ship and its passengers first. The following day the hoaxer was located and arrested and later sentenced to twenty years in jail.

This bomb scare event formed the basis of the 1976 film *Juggernaut*, starring Omar Sharif, Richard Harris and Antony Hopkins, while the ship used was *Maxim Gorkiy*, given the name *Britannic*. In the film the bombs were not a hoax as two explosives were detonated while the ship was at sea. It is of interest to note that the twin AEG steam turbine *Maxim Gorkiy* was originally built in 1969 as *Hamburg*, the same year as *QE2*. She was the largest liner built in Germany since the Second World War. However, since she had become *Maxim Gorkiy* in 1974, her steam turbines had become increasingly expensive to run and she was scrapped in February 2009. Like all things mechanical and electrical, machinery and moving components can sometimes break down or malfunction, or exhibit a spurious alarm, and the complex systems of *QE2* were no different.

On 30 March 1974 *QE2* left New York for a cruise to Bermuda with 1,648 passengers on board, 800 of whom were senior citizens, together with a crew 1,041 under the command of Captain Peter Jackson. It was advertised as a 'Sun & Fun Pigskin Cruise'. On board were members of professional American football teams, including the New York Giants and Jets, the Baltimore Colts and the Washington Redskins. This formed the basis of the lecture programme for the passengers. Rough seas prevented the pilot from landing. The following day, 31 March, Captain Jackson cancelled the mandatory lifeboat drill and advised the passengers to stay in their cabins as the bad weather and rough seas continued. Many of the technical details included in the following account were recounted by a former Cunard engineer on board at the time.

QE2 in mid-Atlantic en route to the Falklands after she had been called up as a troopship. (Crown Copyright)

Although by the following morning, 1 April, the weather had abated, at 4 a.m., while 250 miles south of Bermuda, the turbines were shut down. At the time the engineers were unaware of what was happening until a tube in the aft boiler failed and chaos reigned when the ship 'blacked out'.

With the ship drifting and on emergency power, it became clear why the tube had failed, but it took some time to discover how oil had entered the system. An alarm on an electronic failsafe probe, designed to detect the presence of oil or other contaminants in *QE2*'s boiler feed water system, had failed. This malfunction had allowed oil-contaminated water to enter the water tubes of the boiler tube walls. The oil spread a film through a number of tubes and prevented the transfer of heat from the furnace to the water. Eventually a 'hot-spot' had formed and a blister developed that fractured the tube. The turbines were stopped before any further damage occurred.

The Chief Engineer was instructed by the New York office to proceed to Bermuda and this was attempted by steaming on the starboard boiler only and sailing 'slow ahead'. The affected boiler as well as the whole feed system was still badly contaminated with oil and a few hours later a tube in this boiler also failed. Power on emergency generators was, however, restored.

At the suggestion of Captain Jackson, the following day the Norwegian cruise ship *Sea Venture* set sail from Bermuda to assist *QE2*. Upon *Sea Venture*'s arrival, between 7.30 a.m. and 3.39 p.m. a total of 1,654 persons (including six crew) were transferred to her in calm seas, after which she sailed for Bermuda for their onward return to New York. Two Moran Towing tugs were dispatched from New York to render assistance. Meanwhile, a team of specialists from Houseman & Thompson, experts in oil dispersant and chemical cleaning, had been flown out to Bermuda and came out to *QE2* on the ocean-going tug *Bermudan* together with 10 tons of dispersant and 11 tons of replacement boiler tubes, along with more diesel oil for the emergency generators. They immediately set to work on flushing the one remaining undamaged boiler and the entire feed system. *QE2*'s engineers assisted the Houseman personnel and also opened up the access doors to the two damaged boilers for inspection.

On the after boiler there was one burst furnace tube but the rest looked serviceable. On the second boiler to fail, the starboard one, it was found that one tube had burst, with a majority of the remaining tubes badly blistered almost to the point of bursting. Not long after the Moran tugs arrived to start the tow, the engineers managed to get the port boiler and the rest of the feed system clean enough to flash up. Once steam had been raised, the main turbo-alternator was restarted and the turbines run up to 'slow ahead' to assist the tow of the tugs.

Once *QE2* had arrived at Bermuda, Foster-Wheeler boilermakers came aboard and set about re-tubing the starboard boiler furnace and the burst tube in the after boiler. Once the latter had been repaired, Houseman's started chemically purging it. After a couple of days,

The troopship *QE2* returned to Southampton on 12 June 1982 on the completion of her duties during the Falklands campaign. Red Funnel's hydrofoil *Shearwater 5* is passing her in the right foreground. (Peter A. Alford)

Left and below: Two views showing *QE2* mid paint change after Cunard decided to return her to the Line's traditional hull colour of dark charcoal grey. (Richard de Kerbrech)

the after boiler was considered clean enough and was flashed up and paralleled with the port boiler. With two boilers working, *QE2* left Bermuda for New York on 11 April at 16 knots. Todd's Shipyard of New Jersey then replaced thirty of the 3,000 tubes so that on 16 April she was able to leave New York for Southampton on two boilers, with No. 3 (starboard) still out of action as it had suffered the worst damage by oil contamination.

Upon her arrival in the UK, repairs over the next four days were scheduled around the clock. Engineers pulled out all the stops and 1,000 gallons of trichloroethylene were pumped around the boiler, this process expedited by an access hole being cut in the hull to allow a 2in pipe to be connected from the quayside to the boiler. No. 3 boiler was repaired, cleaned and tested in time for her scheduled twelve-day Mediterranean Cruise on 24 April. During this cruise the engineers managed to restore the ship to full power while en route from Southampton to Naples, making a run of 31 knots. Boiler making, especially the water-tube type as on *QE2*, was a specialist trade with assembly of the thousands of tubes of a curvature shape originally built in a boiler shop. Maintenance and re-tubing of boilers inside boilers while in situ on a ship was even more difficult, the manoeuvring and installation requiring specialist knowledge and expert skills.

In this case, the breakdown, emergency repairs and refund of fares cost Cunard £750,000, which effectively wiped out Cunard's profit for that year.

The subsequent report on the breakdown surmised that 'there was a mechanical failure compounded by human error and a fault built into the plant'. Essentially, the high-pressure fuel pumps that supplied oil to the boilers were each fitted with a steam heating jacket. One of these pumps was dismantled to replace the pumping unit with a spare. Beneath the pump were two drain connections, one from the water jacket and one from the inner oil side of the pump, normally blanked off with a plug. The threads on both were identical and both connections were removed. During the course of reassembly, these two connections were mistakenly switched over. In this way, when started, the oil was pumped into the steam jacket side, which tracked its way via a drain manifold to the main hotwell, also known as the atmospheric drain tank, while other oil was found venting to the bilges. In this manner the oil entered the boiler feed water system. The recommendations that followed called for additional staff in the engine room, more alarm systems in place and an updated arrangement on the steam and feed water open feed system.

More misfortune was to follow some two years later during July 1976. While at Southampton, prior to sailing, the port HP turbine coupling had been opened up for an inspection, possibly for a Maritime & Coastguard Agency (MCA) survey. The *QE2* left port on 23 July with 1,227 passengers aboard. The next morning, around 4 a.m. while 80 miles off the Scilly Isles, the flexible coupling drive connecting the port HP turbine and the reduction gearbox burst apart with parts breaking through the cast iron casing, described as a mighty 'bang'. The coupling was lubricated by free-flowing lube oil, which then spilled out onto the hot superheated pipes and gave off a flammable oil vapour that ignited with a boom! A resulting fireball blew out of the engine room and down the working alleyway. Flames shot up the funnel's ventilation shaft 100ft above, blacking and buckling the funnel casing high above the passenger decks.

As luck would have it, the Second Engineer on duty in the turbine control room, normally manned by three persons, was immediately alerted to the situation and shut down the oil pump. His prompt vigilance and action saved the ship from further damage and the main fire was brought under control after twenty minutes, although with some small insulation fires still burning. A young mechanic was badly burned during this emergency. Meanwhile, a third electrician together with a junior electrician shinned up the twelve decks via the escape stairs to see if anyone was trapped in the engine room lift. While climbing the steel stairs they were rained on by molten aluminium droplets that had melted from the aluminium grating around the steel stairs. Fortunately, they found the lift empty.

Remedial repairs were made whereby the port shaft was hand-jacked round to enable the locking pins to be inserted on the shaft, along with the brake. In the wardroom the rumour was rife that, during the recent inspection, Vosper-Thornycroft staff may have left a spanner in the coupling casing or failed to reassemble it correctly. *QE2* restarted the starboard turbine and returned to Southampton on reduced speed. Upon reaching Cowes Roads she was boarded by Hampshire Constabulary as sabotage had been suspected, but this was not the case.

The report on the fire stated that the damage was the result of a broken coupling, a seizure that resulted due to the misalignment of the turbine rotor and gearbox outside their normal running tolerances. The coupling concerned was the same component as used on P&O's *Canberra* and P&O had one in stock with which *QE2* was repaired. After round-the-clock repair work by engineers the *QE2* sailed on 3 August, making a six-day Atlantic crossing. The departure was during the Cowes Week Regatta and the recent fire incident was made lead feature complete with a photograph in the *Southampton Evening Echo* under the heading 'Move over Cowes, the Big Girl's back in action'. This conveyed a warm, positive note at the time when Cunard was struggling financially.

During October 1974 *QE2*'s main competitor, the French Line's *France* (see Chapter 7) was withdrawn from service after losing £10 million a year. By contrast, by 1978 in a turnaround of fortunes, *QE2* was generating an annual turnover of over £30 million, while her fuel cost £5 million per year and the monthly manning cost for her 950 crew was £225,000. In addition, she cost £80,000 a day to keep idle in port and broke even at 65 per cent occupancy. Trafalgar House learned quickly that *QE2* only generated a profit when at sea full of passengers. For her original Atlantic-shuttle ferry voyages she was scheduled to cross with one and two-day stopovers at each terminal port when essential boiler maintenance could be undertaken. With a more demanding schedule of steaming without any days off, the three boilers were at the maximum of their demand for steam with much limited maintenance. The consequences of the design decision not to install a fourth boiler were being felt.

On 19 March 1982 Argentine forces occupied the island of South Georgia, a former whaling station, and two weeks later they invaded the Falkland Islands, a self-governing British Overseas Territory. A task force was hastily assembled code-named Operation Corporate. *QE2* arrived at Southampton in the early hours of 4 May and along with other merchant vessels became one of the Ships Taken up from Trade (STUFT) to participate. Two days later urgent preparations were made to convert her into a troopship, as indeed the two *Queens* had been in a previous conflict. Furniture, sculptures, paintings, ship models and fruit machines were removed to shore along with stores. Carpets that could not be removed were covered over by hardboard sheeting. Of the ship's permanent crew, 650 had volunteered to sail to the South Atlantic into the war zone. In the meantime, around-the-clock work by Vosper-Thornycroft was under way converting the super-liner into a troopship. The structure of the upper aluminium alloy passenger cabins was not sufficiently strong to take the weight of helipads, so the highest steel deck was chosen to take the feet of the 'I' beam girder supports (similar to RSJs). Prefabricated helipads were built in Vosper-Thornycroft's plate shop and these were hoisted aboard *QE2* and welded in place. She embarked some 6,000 troops in total from 5 Infantry Brigade and a Gurkha regiment.

The ship prepared to sail on 12 May under the command of Captain Peter Jackson but at 2 p.m. it was revealed that one boiler had been shut down for maintenance and a second had developed a massive leak in its distilled water supply. She sailed at 4.03 p.m. with only one boiler fully operational and so the ship of peace went to war. The following day the source of the leak was discovered, a valve that had been left open, so by Friday, 14 May it was necessary to bring the third boiler on stream. The passage south was via Freetown, Sierra Leone, for fuel and water and Ascension Island for further stores. As far as possible the ship was 'darkened' (the defensive process of ensuring that no light whatsoever emanates from a ship to avoid detection at night). The call at Freetown was on 18 May, where 1,867 tons of fuel oil were taken on in three hours.

She then sailed to within 50 miles of Ascension, where helicopters air lifted stores and personnel between the island and the ship. On 27 May *QE2* met other Task Force vessels in the cold, misty, ice berg-ridden waters of South Georgia and transferred her troops by smaller trawlers to *Canberra*. In return, survivors from the attacked destroyers HMS *Antelope*, *Ardent* and *Coventry* were taken on board *QE2* and, prior to her return voyage to Southampton on 1 June, she took on 4,000 tons of heavy fuel oil (HFO) transferred to her from RFA *Bayleaf*. After steaming 14,967 miles and consuming 10,287 tons of fuel oil, *QE2* arrived back at Southampton on 11 June. After first being greeted by RY *Britannia* in The Solent, she received a rapturous welcome in her home port.

The work on converting her into a troopship had taken eight days and it took nine weeks to reinstate her for civilian passenger service. In this her familiar charcoal grey hull was painted over in a 'confederate' grey, but her funnel was painted in traditional Cunard livery. She returned to service on 14 August 1982.

It has been said of Wembley Stadium in London that the venue itself transcended any event or attraction that was held therein. And so it was for *QE2*, for some were attracted by the cruise itineraries offered, others by crossing in her to or from the US. For many, including the authors, it was the appeal of the ship herself, no matter what the accommodation, décor or owner. Some of the factors that ensured her good sea-keeping qualities were her fine hull form with long bilge keels and bulbous bow, her two sets of two retractable Denny-Brown stabilisers, a length to beam ratio of 9.17:1 and a GM of 5ft.

From the breakdown of the turbines during the full-scale trials to the Canaries to a breakdown off Falmouth just after her return to service after the Falklands, *QE2*'s turbines and boilers had proved expensive to maintain. Many of the spares required had to be specially manufactured by hand and this costly maintenance, compounded with the meteoric increase in oil fuel prices since the ship was built and high manning costs caused Trafalgar House to ponder whether to scrap the ship or re-engine her in the hope that she could continue in service for another twenty years.

QE2 at speed in her Confederate grey livery. (Cunard)

Once it was decided that she should be retained, tenders were received for the conversion work and Cunard placed the order with Germany's Lloyd Werft in Bremerhaven. On 27 October 1986 she was drydocked at the shipyard, and during the winter of 1986–87 *QE2* underwent 'heavy surgery' in which her original steam plant was removed along with the propellers. This monumental task involved stripping out the turbines and steam pipe lagging, which contained asbestos. In all, some 4,700 tons of redundant machinery was deposited on the quayside as scrap.

QE2 was re-engined with diesel-electric machinery consisting of nine 9-cylinder, 4-stroke MAN-B&W L58/64 diesel engines built in Augsburg, Germany, installed as the generator prime-movers, coupled to nine water-cooled, salient pole, three-phase synchronous

GEC generators, each of which developed 10.5 MW (14,750 hp) of electric power at 10,000 volts. Each engine weighed around 120 tons and besides supplying ship's auxiliary machinery and hotel services requirements via transformers, the total electrical power they generated drove two GEC main propulsion motors, one on each propeller shaft. The power output from each motor was 44 MW (59,000 hp), which was capable of giving *QE2* a top speed in excess of 32 knots. The motors were of salient pole construction, 29ft 6in in diameter and, weighing 400 tons each, they were the largest marine motors ever built. When running at 144 rev/min the service speed of 28.5 knots was maintained using seven generators or 85 per cent full power so that essential and routine maintenance could be carried out at sea without affecting schedules. Her previous turbines at 28.5 knots had burnt 550 tons of heavy fuel oil (Grade IF80) a day or 120,000 tons per year (at £100 per ton). The consumption of the new plant was around 270 tons per day.

Manoeuvred by tugs, *QE2* at Lloyd Werft on the conclusion of her diesel-electric conversion. (Lloyd Werft)

QE2 in her final aspect outward bound, passing the Needles, Isle of Wight. (Cunard)

The diesel exhausts heated nine heat-recovery boilers coupled with two oil-fired boilers for auxiliaries and domestic fresh water heating. Four flash evaporators and a reverse osmosis plant enabled the engine room to produce 1,400 tons of fresh water daily from sea water.

The 120,000 shp output was required to drive two 19ft, five-bladed, highly skewed, controllable pitch propellers manufactured by LIPS of Drunen in the Netherlands. By altering the pitch of the blades the liner could be manoeuvred astern with the uni-directional motors. In this manner the dedicated astern machinery that had been required for the original fixed-bladed screws could be dispensed with. Abaft of the propellers on each shaft were 28ft diameter 'Grim wheels', each consisting of seven vanes. Their purpose was to absorb the astern thrust component from each screw and convert it via its vanes to an augmenting propeller. It was claimed that this saved 4 per cent of the fuel costs. A porthole was cut in the stern of the ship to observe the performance of the Grim wheels and any propeller cavitation.

In addition to the power plant, the 'hotel' side of the ship underwent many changes, with the Tables of the World Restaurant becoming the Mauretania Restaurant and the Double Down Room rearranged as the Grand Lounge. Following sixteen days of trials, and after six months of work at a cost of £92 million, *QE2*'s refit was completed on 25 April 1987 and the 'transplant' was complete. One of the most striking features was her broader, stockier funnel design, complete with Cunard livery. She sailed from Southampton on 27 April on her post-conversion voyage and arrived in New York on 4 May.

One of the unforeseen consequences of the change from steam to diesel was the effect it was to have on the engine room staff as manning a new 'motor ship' required engineers with 1st and 2nd Class Motor Certificates of Competency. Most of the senior engineers had 1st Class Combined Steam and Motor Certificates, especially those of a younger age. Some older seniors who only had Steam Chief's Certificates, also had 2nd Class Motor. Of these, some went back to college to obtain a motor endorsement. One or two did not have sufficient motor watchkeeping time and had to drop to second or even third engineer to get their qualifying sea time in. Some found the conversion too challenging and retired. The top jobs went to the chief engineers from *Cunard Princess* and *Cunard Countess*, both of which were motorships.

During a guarantee refit *QE2* was drydocked in the King George V Graving Dock at Southampton over 9–10 July 1987. The Grim wheel vanes had broken off down to the stubs on the hubs but it was not fully known when they sheared off. Some said it was while going down the river from Bremerhaven, others thought they had broken off when going from full ahead to full astern during the trials. The stern porthole was welded over in the guarantee refit and the Grim wheel hubs removed. Whether they had in fact increased the propulsive efficiency of the ship or saved on fuel consumption remained inconclusive!

In later years the passenger accommodation of *QE2* was upgraded significantly and augmented, and it included new facilities such as a full health spa. Sometimes the planning and execution of subsequent refits fell short of the company's expectations and the press often had a field day with reports of overflowing toilets, flooded cabins and lack of air conditioning as the ship re-entered service. By the 1990s it was very much Cunard's heritage and tradition that sold berths, benefitting *QE2* with a large renaissance in her patronage.

One of the unforeseen incidents that occurred and removed her from service was a grounding on an uncharted shoal of rocks off Martha's Vineyard when a late change in her original cruise schedule caused a change to *QE2*'s itinerary. She left New York on 3 August 1992 with 1,824 passengers and at 9.58 p.m. on 7 August, while off Cuttyhunk Island, a sudden vibration was felt in the ship, compelling her to be brought to a standstill. The engine control room confirmed that motors and propeller shafts were still operational with the controllable pitch screws altered to zero pitch. As a collision was ruled out, the outcome was that the liner had in fact grounded. The initial damage assessment revealed that a sounding pipe to No. 1 saltwater double bottom tank was fractured with sea water entering.

The passengers were disembarked to nearby Newport, Rhode Island, in the ship's lifeboats and aboard two local ferries. *QE2* then headed for the Boston Navy Yard for drydocking and an assessment of the underwater damage to the hull. High and dry in the graving dock, the damage was revealed to be extensive, running some 400ft aft from the bulbous bow and over a width of 80ft over the keel. Moreover, either side of the keel were indents, some 240ft in length and 14in deep, plus scars and gouges. Twenty inner bottom tanks, mostly empty or containing fresh water, had been damaged but No. 10, an empty fuel tank, was the only tank to have leaked a small amount of residue. In addition, 32ft of the port bilge keel had been damaged during the grounding.

Only temporary repairs were carried out because the correct specification of steel was unavailable in the navy yard. Therefore, the decision was made to transfer her to Blohm & Voss in Hamburg. She proceeded across the Atlantic at 24 knots and upon arrival there

A tight squeeze as *QE2* passes through one of the Panama Canal locks. Along with Norddeutscher Lloyd's *Bremen*, she was the only other super-liner to pass through the canal. (Cunard)

QE2 in the River Mersey on 24 July 1990. She had sailed into Liverpool to celebrate the 150th Anniversary of the Cunard Line's association with the Port of Liverpool. (Philip Parker)

Following her sale to Nakheel of the United Arab Emirates, *QE2* was drydocked in Dubai prior to conversion into the Hotel QE2. (Drydocks World)

more permanent repairs were undertaken in the huge floating dock. The damaged steel hull plates were replaced by new ones, welded in place. One former Chief Engineer was given a removed piece of the hull with the gouges across it as a souvenir. It was chrome plated and engraved with the date of the grounding.

One of the unforeseen benefits of this incident was that metallurgical macro and microscopic examinations of the ship's 24-year-old hull plates that had been replaced showed that the steel was in remarkably good condition. If this was extrapolated across the whole hull it meant that *QE2* would have a longer service life ahead of her.

The reason for the grounding was an effect called 'squat', whereby a vessel under way, moving through shallow water, displaces more water than would be the case in a deeper sea. As speed is gained, the stern pulls down and when *QE2* was travelling at 25 knots it was estimated she had a squat of 6ft 6in. In this case she had run out of water over the shoals and grounded on the highest pinnacles. Several of her cruises had to be cancelled, and the repairs in Boston and Hamburg, the passenger refunds on seven scheduled cruises plus manning wages, cost the company around £30 million. On completion of the repairs, *QE2* returned to Southampton on 3 October to a 'maiden voyage'-style welcome.

QE2 berthed at Port Rashid, Dubai, on 13 October 2022, converted into a luxury hotel, beginning a new phase of her long career in semi-retirement in the sun. (Hotel QE2 Dubai)

Two years later, on 14 November 1994, *QE2* was taken out of passenger service in New York and sailed directly for Hamburg, with contractors on board preparing for her major 'Project Lifestyle' refit, with the emphasis on the ship's and Cunard Line's history. In this her gross tonnage increased to 70,327.

Trafalgar House was acquired by the Norwegian Kvaerner group in 1996 and they subsequently sold Cunard to the Carnival Corporation for $630 million. Moving into the twenty-first century, on 29 August 2002 she became the first merchant ship to sail 5 million nautical miles. When *Queen Mary 2* entered service in 2004, *QE2* reverted to full-time cruising and on 5 November that year she became Cunard's longest-serving express liner, surpassing the earlier *Aquitania* (1914–50).

However, after forty years of service, she had become too expensive to maintain in top condition and she was withdrawn from service in October 2008 and sold to Nakheel of Dubai for $100 million, for use as a floating hotel. In her time she had completed some 803 Atlantic crossings, sailed some 5.6 million nautical miles and carried in excess of 25 million passengers. On 11 November she commenced her final voyage with an emotional farewell firework display, departing from Southampton surrounded by a flotilla of small boats. She arrived at Port Rashid in Dubai on 26 November 2008 and was handed over to her new owners the following day. Since then, after a slow start she has undergone conversion to a five-star plus hotel and convention centre. *QE2* was a classic super-liner, an icon, a ship for all oceans and a national treasure. Now that she is retired in a foreign land, it is to be hoped that her current owners will ensure her future.

NINE

A SUPER-LINER FOR A NEW MILLENNIUM: THE EPIC *QUEEN MARY 2*

The genesis of *Queen Mary 2*, the greatest transatlantic super-liner ever built, followed along a similar path to that of the earlier *Queen Elizabeth 2*. In both cases an alternative option had been considered before the final design was settled on. Where *Queen Elizabeth 2* was concerned, she had been preceded by the ill-advised *Q3* concept for a traditional three-class liner dedicated to all-year-round employment in the Atlantic trade. Also, prior to the *Queen Mary 2* project, an earlier proposal had emerged. Not for a vessel to undertake a mode of operation at variance to that which she went on to follow, since that proposed format had already been established with the *QE2* as a blend of transatlantic scheduled service and off-season cruising, but rather in the form of a very different design to that which ultimately resulted.

The ship design in question, dubbed *Q5*, was intended as the initial replacement visualised for *QE2*, which at the time was 22 years old. Even with her replacement diesel-electric engines installed during her 1987 life-extending refit, she would be nearing the end of her life in just eight years, or so it was thought back then, and it would take much of that time to design and build her successor.

Since 1971 Cunard had been owned by Trafalgar House Investments but, twenty years later, the company was not in good financial shape having invested in a range of business activities, too many of which were producing poor results. Nonetheless, Trafalgar House regarded Cunard as the jewel in its crown and was keen to sustain the line's position as a major player in the cruise market and, singularly, as the sole continuing transatlantic ocean travel option. Engaging Deltamarin's top naval architect, Njal Eide, the 'Century 21' project was unveiled in December 1991, otherwise referred to as *Q5*. Various configurations were revealed in the preliminary design documents, with one, two or three funnels, for a ship of 90,000 gross tons with a length of 985ft overall (later increased to 1,045ft) and a beam of 128ft. Like *QE2*, she would have had diesel-electric engines, although steam turbine propulsion was also considered but, extraordinarily, a service speed of 37.5 knots on the Atlantic run was intended. This flew in the face of the reasoning behind the fuel-saving change made to *QE2*'s engines but the philosophy advanced was that charging each passenger $5 extra per day on a crossing of shorter duration made better economics than reducing bunker costs for the voyage by 10 per cent.

All this proved to be academic, for the project was not pursued, and six years later Cunard was acquired by the Norwegian company Kvaerner Group. Just two years after that, in 1998, amidst continuing rumblings about an unspecified *QE2* replacement, Cunard was sold again to the Carnival Corporation. These new owners wasted no time in turning talk into action, first with designs for two new 84,000 gross ton ships for Cunard and then, when that idea was abandoned, the 'Project Queen Mary' was announced on 8 June 1998 for what was to become the first new liner for transatlantic service since the 1960s. Not only that, she would surpass everything that had preceded her by an astounding margin. Briefly the largest passenger ship in the world when she entered service on 12 January 2004, her tale of the tape revealed her to be a prodigious vessel, far bigger than *Q5*, having a gross tonnage of 148,528 and almost double that of the previous largest super-liner *Queen Elizabeth*, which had measured 83,673 gross tons.

Queen Mary 2 under construction at the Chantiers de l'Atlantique shipyard in a drydock view before her funnel was fitted. The base structure was fitted on 13 March 2003 and the topmost section on 4 April. (Paul A. Massey)

Fitting out of *Queen Mary 2* in Bassin C at the Chantiers de l'Atlantique, was well advanced by 7 July 2003 when this photograph was taken. (Maurizio Eliseo)

The design of the new Cunarder was entrusted to Stephen Payne, a young naval architect. His task was to create a new-technology ship yet to evoke the spirit of the golden age of Atlantic travel while combining in a single ship the requirements for line service duties for eight to nine months of the year with an annual cruise programme capability. This involved raising the passenger accommodation from a low position within the hull in order to provide a high percentage of outside and balconied staterooms, while simultaneously retaining the popular wrap-around open boat deck to allow passengers to take the traditional open-air promenade. For the latter, an exemption was granted in the design because of *Queen Mary 2*'s North Atlantic service, allowing lifeboat stowage to be 88ft above the water rather than tucked inside the hull at the SOLAS-stipulated 52ft 6in, as on cruise ships. Other design stipulations required her to have a block coefficient of 0.61 to enhance stability, and a fatigue life of forty years based on 60 per cent active utilisation. It is believed that this was the first time ever that a fatigue life was specified for an ocean liner. Another requirement was for thicker plate scantlings, in certain areas as thick as 28mm, and the adoption of revised welding and grinding techniques in order to avoid any weaknesses concealed between the plates and to strengthen the plate bonds.

The forward superstructure was to be set back and curved in profile to minimise wave damage in heavy weather, and the foredeck would feature a whale-back and breakwater for water dispersal. Her bow was to have an elegant curved sweep to the stem continuing downward to an extended bulbous bow. At the aft end there would be a Constanzi-style stern, combining a rounded spoon shape set within a flat transom.

To enhance superstructure aerodynamics plus the clearance of funnel exhausts, and even to assess open deck wind impact on passengers, wind tunnel testing was performed by FORCE Technology Division for Maritime Industry at Lyngby, north of Copenhagen, Denmark. As the culmination of the preliminary design effort, the General Arrangement plans for the new ship were completed and approved by 8 November 1999.

Queen Mary 2, as large as she is, was originally intended to be even bigger, but operational considerations necessitated restrictions to her dimensions, notably to provide adequate clearance for her to pass under the Verrazano Narrows Bridge in the approach to Manhattan. Even so, at high tide, the clearance is a scant 10ft.

Tenders for the construction of *Queen Mary 2* were invited in December 1999 from Harland & Wolff, Aker Kvaerner, Fincantieri, Meyerwerft and the Alsthom Marine subsidiary, Chantiers de l'Atlantique. Stephen Payne was to comment, when these bids were being evaluated, that some of the building yards seemed to be only interested in cruise ship construction, a less onerous undertaking. The decision of Carnival's design organisation was to award the contract for the job to the Chantiers de l'Atlantique at Saint-Nazaire, the birthplace of *Normandie* and *France*, and a letter of intent to that effect was signed with the yard on 10 March 2000. Formal signing of the contract, now designated as Yard No. G32, took place in Paris on 6 November 2000 between Micky Arison for the Carnival Corporation and Patrick Boissier, the President and CEO of Alsthom Marine.

Naturally enough, doubt was expressed in the trade press as to whether the new Cunard ship would be anything more than a glorified cruise ship but, in answer to the sceptics, a spokesperson for Chantiers de l'Atlantique made it indelibly clear: 'The *Queen Mary 2* will not be a cruise ship. We are building a liner!'

First steel for Cunard's new liner was cut on 16 January 2002 and her keel was laid six months later, on 4 July, in the Construction Dock B at Saint-Nazaire with its co-located pre-assembly area. In total, some ninety-four structural blocks, welded together from 300,000 pieces of steel, were transferred into the drydock for assembly into the hull and superstructure. For a ship of her size, building progress was expeditious and she was ready to be floated in her construction dock on 21 March 2003. Fitting out ensued in Bassin C with equal promptitude, and preliminary shipyard sea trials were conducted between 25 and 29 September 2003. She was evidently not complete at this stage, only partially painted and lacking the

Queen Mary 2 nears completion in this fitting-out dock view. (Alstom Marine)

Queen Mary 2 during her first series of sea trials. The paintwork remains to be completed and at this date she doesn't have her full complement of lifeboats. (Alstom Marine)

Maiden arrival of *Queen Mary 2* at Southampton on 26 December 2003. (David L. Williams)

majority of her lifeboats. During the high-speed trials, one of her bow thruster doors was forced off its hinges, tearing part of the surrounding hull plate. Back at Saint-Nazaire, a replacement could not be obtained promptly enough from the supplier so one was prefabricated by the shipyard under the guidance of Jean-Jacques Catepaille, the senior designer.

After she had returned to the shipyard for the final stages of fitting out, sea trials were resumed on behalf of the owners with her designated master, Captain Ronald Warwick (he became Commodore Warwick on 26 December 2003), in command. They were completed satisfactorily between 7 and 11 November, running between Saint-Nazaire and the islands of Île d'Yeu and Belle-Île where, over a 10-mile distance, she consistently achieved her design maximum speed, reaching an average of 29.63 knots. Back in the drydock, tragedy struck, marring the propitious omens for *Queen Mary 2*, when a gangway collapsed on 15 November during an open-day visit for shipyard workers and their families. A total of forty-eight persons plunged more than 50ft to the base of the drydock, sixteen of them fatally injured.

Following that tragic incident, which inevitably cast a cloud over what should have been a momentous time for both the ship and the yard, the completed *Queen Mary 2* left Saint-Nazaire for Southampton, where her inaugural arrival on 26 December 2003 was met by an ecstatic welcome despite the chill of a cold winter's day. Her owners formally took delivery and fifteen days later, on 8 January 2004, she was christened by Her Majesty Queen Elizabeth II. It was the second *Queen* liner upon which this honour had been bestowed and the third naming ceremony of a *Queen* attended by Queen Elizabeth. Previously, she had been present at the launch of the liner *Queen Elizabeth* in 1938 and she had personally named *Queen Elizabeth 2* in September 1967.

After all the expected pomp and circumstance, *Queen Mary 2* set off across the Atlantic on her maiden voyage, departing Southampton on 12 January 2004, not destined for New York but to Fort Lauderdale, Florida, after calling at Funchal, Las Palmas and Barbados. The occasion was marked by great fanfare as the new *Queen* repeatedly sounded her horns, an audible reminder of her former namesake, having one of the steam whistles first installed on *Queen Mary* mounted upon the portside of her funnel. It had been specially reconditioned by Kockums AB of Sweden, who had manufactured it originally back in 1936.

Another view of *Queen Mary 2* on her maiden arrival at Southampton, after she had turned about off the Western Docks and was heading towards the Queen Elizabeth II passenger terminal. Her imposing size and grandeur are evident in this view. (David L. Williams)

Not only was *Queen Mary 2* the largest passenger ship in the world and, in 2003, ever built, but she was also the longest at 1,132ft overall and the broadest with a beam of 148ft. Her draught was 33ft. She had cost $780 million to build, a figure that ranks with the $840 million cost of *Normandie* when 1935 values are escalated to 2000 levels. As completed, her passenger capacity on the lower berth basis was 2,620, increasing to 3,090 with upper berths included. That compares to circa 5,250 on a cruise ship of comparable tonnage. Her crew numbers were 1,253, making a total complement of around 3,850. The accommodation was in a single class with virtually the complete run of the ship to all, although there were certain on-board distinctions for those travelling in what was designated as Grill standard.

If the size and proportions of *Queen Mary 2* were extraordinary, so too were the public spaces aboard her, her vast interior capacity and area making possible an overall fusion the like of which has never before been seen on a super-liner. Like earlier super-liners, the fumes from her diesel engines were exhausted through split vents taken up the sides of the hull and reconnected beneath the funnel. This permitted the exploitation of a wide-open space through the centre of the ship, with walkway access connecting all the main public rooms.

Queen Mary 2 outbound through The Solent on 16 June 2007 on a New York crossing. (David L. Williams)

As first completed she had two large theatres, the Royal Court seating 1,094 persons, and the cinema-style Illuminations with 493 seats, doubling as a planetarium. She had two large restaurants: the Britannia arranged over three deck levels, one of the largest ever built into a passenger liner and capable of serving 1,100 passengers at each of two sittings, and the King's Court lido-style restaurant able to quarter into four different, themed dining areas. Besides these, there were two à la carte grill restaurants, the Princess and the Queens, the former catering for the 178 passengers in deluxe staterooms, the latter for the 200 passengers occupying suites, penthouse and duplex staterooms. And so it went on with bars, ballrooms and lounges. As a whole they reflected her dual-role status. Apart from her many varied public rooms, there was also a plentiful provision of exterior deck space and recreational amenities, more than on any previous super-liner. For both route service and cruising, it suggested a luxury spa hotel. The décor consciously revived the unique Cunard take on art deco, reflecting the era of the first two *Queens* in which it was first introduced.

In keeping with her distinctive, in some respects revolutionary, design parameters, the power plant and propulsion system of *Queen Mary 2* deviates entirely from the steam turbine engines of the past and even the diesel-electric drive introduced on *Queen Elizabeth 2* during her major refit in 1987. The new Cunarder's engines are complex but very efficient, a combination of two prime movers designed to provide the best performance for her dual mode of operation, providing in addition an appreciable supply of power for the ship's hotel services.

They comprise two General Electric Marine LM2500+ aeroderivative gas turbines, each of 25 MW power output, and four Wartsila 16V46 Enviro Engine diesels, each producing 16.8 MW and coupled to ABB Finland alternators. The complete system can generate a total of 117.2 MW (157,000 hp). While on transatlantic service, having a surplus of power available to call upon in adverse weather conditions, *Queen Mary 2* deploys her entire power system, but when cruising she relies on power generation from the diesel-electric engines alone. The machinery selection has permitted greater efficiency in the utilisation of hull space, with the result that it occupies a fraction of the space taken up by, for instance, the steam turbine engines of the original *Queen Mary*. The diesel-electric power sets are located in two adjacent engine rooms, whereas the gas turbine generators are mounted high up in the funnel casing, obviating the requirement for additional engine room space.

The propulsion arrangement consists of four Rolls-Royce AB-Alstom Kamewa MW Mermaid variable-speed, reversible-motor thruster pods, mounted in a tractor or pulling configuration. Two of them are of azimuth, steerable type, the other pair forward-mounted fixed units. Together the pods provide a propulsive power of 86 MW. The disposition of the pods was determined during hull model tests in February 2001 at the Wageningen laboratories of the Dutch research institute MARIN, as part of a suite of hydrodynamic optimisation studies. Fitted to each pod is a four-bladed, 19ft 4in diameter propeller. Not only was this entire configuration unique but *Queen Mary 2* was the first vessel in the world to be driven by an agglomeration of four pods. She does not have a conventional rudder, all steering dependent on the two steerable Azipods.

While this power and propulsion system gives *Queen Mary 2* a rated maximum speed of 29 knots, although on her trials she had demonstrated the capability of exceeding that by a small margin, her propulsion pods have not been without their teething problems.

There had been an issue with one of the pods at the time of her floating in drydock in March 2003, delaying the event by four days. Vibration caused by bearing wear had been encountered during test runs in the dock. Subsequently, a sequence of troubles and breakdowns of the pods caused interruptions to her programmed itineraries and both inconvenience and irritation for passengers.

Compared with conventional propellers on shafts, the Mermaid pods are rather exposed beneath the hull and, as if to highlight the point, one of the pods was damaged on 19 January 2006 after striking the side of the deepwater channel at Fort Lauderdale as *Queen Mary 2* was departing Port Everglades on a thirty-eight-day round South America cruise. The departure was delayed and a number of the scheduled ports of call were cancelled because of reduced speed. The passengers protested vehemently and the affair cost Cunard a great deal in compensation, but it also added up to a huge dent in the ship's reputation. *Queen Mary 2*'s subsequent programme was disrupted while she awaited repairs as she had to continue to operate on three pods for much of the remainder of that season. In June, the damaged pod was removed after entering drydock in Germany and then, in November 2006, on a second visit to the Blohm & Voss shipyard the repaired pod was reinstalled.

Another view in The Solent, passing Cowes, Isle of Wight. This stern view shows *Queen Mary 2*'s unique Constanzi stern configuration. (David L. Williams)

However, the propeller pods remained prone to failure, attributed to excessive wear to the thrust bearings of the pod motors. Eventually this led Cunard, represented by Carnival, to resort to litigation by suing Rolls-Royce in the United States' courts, alleging inherent design defects that, it was claimed, the pod manufacturers had been aware of all along and which they had deliberately concealed in order to secure the supply contract. The court found in favour of Cunard and Carnival, awarding damages of $24 million (£15 million). It has to be said that the pods issue has settled down in the ensuing years and operation of *Queen Mary 2* has since been largely trouble free in respect of the propulsion motors.

Queen Mary 2 also has three bow thruster sets, each of 3.2 MW power, and her operational equipment includes four Brown Brothers VM series folding-fin stabilisers with butterfly valve doors. Designed and manufactured by Rolls-Royce in the UK, they are capable of roll attenuation of up to 90 per cent when deployed in combination.

On 16 April 2004, *Queen Mary 2* took over entirely the transatlantic duties of *Queen Elizabeth 2*, beginning with her maiden transatlantic crossing on the New York run. Nine days later, on 25 April, *Queen Mary 2* and *Queen Elizabeth 2* left New York together and sailed across the Atlantic in tandem, the first time that two Cunard super-liners had made the passage in partnership. That same year, on 1 May 2004, the status of flagship of the Cunard fleet was also transferred to *Queen Mary 2*.

Queen Mary 2 seen passing under the Verrazano Narrows Bridge, New York, giving an indication of how small the clearance is beneath the bridge. For passing road traffic she must make a striking impression. (Bill Miller collection)

Since her inauguration, she has clocked up some impressive figures. In her full first year she made twenty transatlantic crossings (equal to ten round trips), thereafter maintaining that scheduled frequency year after year, apart from during the Covid pandemic interruption. She commenced her first world cruise on 10 January 2007, circling the globe in eighty-one days, a feature of her annual programme that has been repeated every year since. By her tenth anniversary she had completed 419 voyages, accumulated 1.5 million sea miles and had amassed a passenger total of 1.3 million. The year before, commencing 6 July 2013, witnessed her 200th transatlantic crossing, her passenger numbers on the Southampton–New York–Southampton run having by then reached 500,000. Her 400th North Atlantic crossing commenced on 23 August 2024.

Despite her evident success, however, the profitability of *Queen Mary 2* is potentially marginal, her comparatively low passenger capacity for a vessel of her size limiting her income relative to her operating costs. In 2016, steps were taken to enhance her earning power in conjunction with a major refit in which significant structural modifications were incorporated.

She was taken in hand on 26 May 2016 in the Elbe No. 17 drydock at Blohm & Voss, Hamburg, for a substantial quantity of work, all due for completion less than a month later on 21 June. In fact, much advanced preparation had taken place by the prefabrication of the structural components, following the placement of the £90 million contract and before her arrival. The principal task involved the addition of thirty-five new balcony staterooms on the forward part of deck 13, in an area taken from the Regatta pool and bar, and fifteen new single-occupancy cabins on decks 2 and 3, utilising space taken from the ship's casino and photo gallery. Other work included the revamping of the layout of the King's Court restaurant to ease mealtime congestion problems, also the removal of the glass elevator from the grand lobby. This permitted renovation of the entire reception area, while a completely new lounge was created

In drydock at Hamburg, *Queen Mary 2* during one of her refits. (Lloyd Werft)

Queen Mary 2 in drydock at Hamburg in 2016 at the time of the major refit that provided an increase to her passenger capacity resulting from a changed profile to the upper deck layout as well as interior modifications. The changes led to an increase to her gross tonnage. (Cunard)

on deck 7 in place of the winter garden. The opportunity was also taken to revitalise furnishings and carpeting throughout the vessel and to implement a further upgrade to the Azipods.

On completion of this sizeable refit, the gross tonnage of *Queen Mary 2* has increased to 151,215. Her revised passenger complement is 2,691 (3,108 maximum with upper berths) and the crew numbers have risen to 1,292. Another big refit followed in the summer of 2023 in the Damen shipyard in Rotterdam, the main focus being overhaul and maintenance of the power and propulsion systems along with other improvements to help elevate passenger experience. Another refit of some magnitude is planned for 2025–26.

* * *

Queen Mary 2 is a truly marvellous super-liner, the total fulfilment of her owners' specified objectives: her structure, unique propulsive power system, stylish accommodation and internal ambience all permeate her design. The result perpetuates Cunard's legacy of internal grandeur, elegance and comfort while satisfying all the perceived demands of a niche market catering for modern-day line travellers and leisure excursionists.

She is an innovative and visionary liner of contemporary styling and she has been highly successful – Cunard, Carnival and Stephen Payne all got their sums right. Although she requires a substantial return on her capital outlay throughout her entire service life, without doubt she is proving to represent a very sound investment. As a modern super-liner she bears little comparison with the earlier *Queens* and other super-liners that pre-date her, but it should be kept in mind that she is a different ship for a different era, facing environmental pressures besides economic pressures, giving fuel efficiency and ecological practices a greater urgency. Also, the expectations of her passengers, across the spectrum, are a world away from those of previous generations.

Without doubt *Queen Mary 2* is a proud member of the exclusive super-liner lineage, although she may well be destined to be the last of that incomparable line, the end of a dynasty.

EPILOGUE

The super-liners' fortunes and fates, governed by the times in which they existed, varied enormously, as did their commercial success or otherwise.

It was a glittering, golden era beset by persistent, unsettling turmoil – military, political and financial. Cunard's *Queens* fared best, true ocean royalty, amassing a collective total of almost 120 years of active service – with more to come! *Queen Mary* survives, despite difficulties, at Long Beach, California, as a memorial to her era, whereas *Queen Elizabeth*, always to some extent in her shadow, had her days ended, in a pointless Valhalla, at the very point of her taking on a worthwhile second role.

Others were lost in funeral pyres that were equally the seed of human folly: *Bremen* destroyed in an act of senseless sabotage unrelated to her wartime circumstances, and *Normandie* through an over-zealous, over-hasty and disorganised, though well-intentioned, determination to prepare her for what would have been a valuable auxiliary role. The fact that transportation of US servicemen on the scale she offered was not required until over a year later compounds the absurdity of the fate inflicted upon her. The incomparable wartime record of the first two *Queens* provides a hint of what *Lafayette*, ex *Normandie*, could also have achieved.

As a monument to a later period of the super-liners, *Queen Elizabeth 2* is also preserved, in her case at Dubai, but considerably altered internally to suit the tastes expected by today's exclusive hotel guests.

Ballin's grand plans for the *Imperator* trio were thwarted by the First World War, the first indication of the impact that this and the second global conflict were to have on the great super-liners. Only three of their number were physically lost during wartime but the course of the futures of another four were altered dramatically, in the case of *Leviathan*, ex *Vaterland*, in a blighted post-war service career epitomised by long spells laid up inactive.

Rex was another of those that suffered the misfortunes of war, sunk when her whereabouts were divulged by Yugoslav partisans who shared no passion for what to them was no more than a symbol of their former Italian aggressors.

United States and *France*, the last of the super-liners to be designed for exclusive, all-year-round transatlantic service, encountered, like the ageing *Queens*, the burgeoning competition of air travel, causing their careers to be ended prematurely when operating subsidies were withdrawn. Yet in truth, their respective conceptions, representing to a significant extent the pursuit of patriotic goals, had failed to see the writing on the wall.

Perhaps, the saddest legacy of the great super-liners has been the protracted abandonment and consequential deterioration of the once magnificent final Atlantic Blue Riband record-breaker. For a country that has always enthusiastically waved the flag and which places a high value on patriotism and national achievement, the treatment of the iconic *United States* can only be regarded as nothing short of lamentable, negligent and even disrespectful.

Queen Mary 2 remains active, already beyond her twentieth season and a highly successful ship in her combined role, a testament to the immense capabilities of these incomparable ships when operated at a less inclement juncture.

Where it began. The Ballin trio of super-liners, as *Majestic*, *Leviathan* and *Berengaria* (right), in the Ocean Dock at Southampton in the 1920s. (Sueddeutsche Zeitung Photo / Alamy Stock Photo)

The three greatest transatlantic super-liners together for the only time, at New York in February 1942. From the left, *Queen Elizabeth*, *Queen Mary* and *Normandie*. (Authors' collection)

Perhaps the last of the great transatlantic super-liners, *Queen Elizabeth 2* (nearest) and *Queen Mary 2*. (PA Images / Alamy Stock Photo)

BIBLIOGRAPHY

Angas, Cdr. W. Mack, *Rivalry on the Atlantic* (Lee Furman, 1939)

Ardman, Harvey, *Normandie, Her Life & Times* (Franklin Watts, 1985)

Armstrong, Warren, *Atlantic Highway* (Harrap, 1961)

Bonsor, Norman R.P., *North Atlantic Seaway* (Stephenson, 1955)

Braynard, Frank O., *Lives of the Liners* (Cornell, 1947)

——, *Leviathan, The World's Greatest Ship* (South Street Seaport, 1973–79)

——, *The Big Ship – The Story of the SS United States* (Turner, 2011)

——, *By Their Works Ye Shall Know Them* (Gibbs & Cox Inc., 1968)

Braynard, Frank O. & Miller, William H., *Fifty Famous Liners* (Patrick Stephens, 1982)

Brouard, Jean-Yves, *Normandie, Le Plus Beau Paquebot de Tous Les Temps* (Navires et Histoire, numero 23); *Liberté, Seigneur de L'Atlantique Nord* (Navires et Histoire, numero 45)

Gaede, George F., *American Super-Liners – They Will Pay* (paper presented to the Society of Naval Architects & Marine Engineers, New York, 1938)

Gallagher, Michael, *The Building of Queen Elizabeth 2* (Ferry Publications, 2017)

Griffiths, Denis, *Power of the Great Liners* (Haynes/PSL, 1990)

Groner, Major Erich, *German Warships, 1815–1945* (Naval Institute Press, 1990)

Harding, Stephen, *Great Liners at War* (Motorbooks International, 1997)

Huchthausen, Peter A., *Shadow Voyage* (Wiley, 2005)

Hutchings, David F., *Pride of the North Atlantic* (Kingfisher, 2003)

Isherwood, J.H., *Steamers of the Past* (Sea Breezes, 1966)

Jung, Ingvar, *The Marine Turbine (Part 2)* (National Maritime Museum, 1986)

Kameen, Tom, *More Than a Century of Machinery Progress in the Cunard Line Passenger Vessels* (The Motor Ship, 1972)

de Kerbrech, Richard P., *Ships of the White Star Line* (Ian Allan, 2009)

Kludas, Arnold, *Great Passenger Ships of the World* (Patrick Stephens, 1975–77)

——, *Record Breakers of the North Atlantic* (Chatham, 2000)

Kohler, Peter, *The Lido Fleet* (Seadragon, 1998)

Mackenzie-Kennedy, C., *The Atlantic Blue Riband* (William Session Ltd, 1993)

Maxtone-Graham, John, *North Atlantic Run* (Cassell, 1972)

Payne, Dr. Stephen, RMS *Queen Mary 2 Owners Workshop Manual* (Haynes, 2013)

Potter, Neil & Frost, Jack, *The Mary* (Harrap, 1961)

——, *Queen Elizabeth* (Harrap, 1969)

——, *QE2: The Authorised Story* (Harrap, 1969)

Rowland, K.T., *Steam at Sea – A History of Steam Navigation* (David & Charles, 1970)

Rushbrook, Frank, *Fire Aboard* (Brown, Son & Ferguson, 1998)

Sims, Philip, *Comparative Naval Architecture of Passenger Ships* (SNAME Transactions Vol. III, 2003)

Spratt, H.P., *Outline History of Transatlantic Steam Navigation* (Science Museum/HMSO, 1950)

Various, *SS Normandie, SS France, SS Norway* (303 Arts, Recherches et Créations, Loire-Atlantique, 2000)

Williams, David L., *Glory Days: Transatlantic Liners* (Ian Allan, 2000)

——, *Cunard's Legendary Queens of the Seas* (Ian Allan, 2004)

Williams, David L. & de Kerbrech, Richard P., *Damned by Destiny* (Teredo, 1982)

——, *Cunard White Star Liners of the 1930s* (Conway, 1988)

——, *Great Passenger Ships That Never Were* (The History Press, 2019)

Wilson, R.M., *The Big Ships* (Cassell, 1956)

REPORTS

Alstom Chantiers de L'Atlantique, *Genesis of a Queen: Cunard Line's Queen Mary 2* (The Naval Architect, 2004)

CIA, US Government, *German Plans for the Invasion of England, 1940, Operation 'SEALION'* (Reading Rooms document 2C.201940)

House of Representatives Committee on Naval Affairs, *Report on the Fire and Capsizing of the USS Lafayette formerly the TEL Normandie* (House Report No. 2031, 20 April 1942)

National Transportation Safety Board, Washington DC, *Boiler Rupture on the SS Norway* (Marine Accident Brief, MAB0703, 25 May 2003)

Project Gutenberg, *EBook of History of the U.S.S. Leviathan* (File 59699)

US Navy Dept Bureau of Ships, *Report on the Salvage of the USS Lafayette ex SS Normandie* (NAVSHIPS-250-880-21, 1946)

The Collision Between HMS Curacoa and RMS Queen Mary (Navigation and Direction Bulletin, December 1949)

Loss of the SS Seawise University (Report of the Marine Court, Hong Kong, 1972)

ACKNOWLEDGEMENTS

We would like to extend our gratitude and thanks to the following persons and organisations for their kindness and support in contributing information and illustrations or in providing other help. Without their assistance, this publication would certainly not have come to fruition.

Our particular thanks go to Jean-Yves Brouard, a world authority on the French Line (CGT) and other French shipping companies, who tragically was killed in a road traffic accident during the writing of this book.

Also, we would single out for particular mention and gratitude Terence Kearney, a former Chief Electrical Engineer who served thirty-nine years with Cunard and twenty-eight on *Queen Elizabeth 2*; also, another former *QE2* Senior Engineer, Martin Harrison.

We would like also to acknowledge the assistance received from Kenneth Beken; Andrew Bennett, a former *Canberra* engineer; Nereo Castelli; Maurizio Eliseo; Michael Gallagher, former PR Manager of the Cunard Line; David Hutchings; the late Arnold Kludas; Bill Miller, 'Mr. Ocean Liner'; Nigel Scutt; Fabien Trehet; Adrian Vicary and Jane Williams, and anyone else whom we have inadvertently overlooked.

Our thanks are also due to Chantiers de l'Atlantique; the Cunard Archives, University of Liverpool; Glasgow University Archives; US National Archives & Records Administration and US Naval Historical & Heritage Command.

And finally, our thanks to Amy Rigg and Chrissy McMorris of The History Press in sharing our vision for this project and for taking it up.

Bow head-on view of *Queen Mary*, a striking photograph of the great super-liner at speed, 'a rampart of a ship … parting the seas in sunder'. (Captain John Trevor Jones/David Hutchings collection)